I0035425

Diabetes? Insulin-dependent?...

Functional Use of Insulin

The advanced diabetes guide for personalized insulin treatment with self-monitoring and multiple daily injections or an insulin pump

English edition based on the 8th German Edition of "Insulinabhängig?...", K. Howorka, Kirchheim Publishers, Mainz, 8th ed. 2009

With a foreword from Professor V. Frankl, MD

Author:
Kinga Howorka, MD
Professor of Internal Medicine
MBA Applied Biomedicine
MPH Master of Public Health
ISO 9001 Quality Certified International Research Group
Functional Rehabilitation and Group Education
Medical University of Vienna, Austria
E-mail: kinga.howorka@meduniwien.ac.at
www.diabetesFIT.org

Written under the cooperation with Jiri Pumprla, MD, MBA, MPH

Graphics:
Karl Alex, Vienna
Mt. Everest views, photos by Geri Winkler (www.winklerworld.net)

ISBN number: 978-1-4461-8753-1

To the real Jonathan Seagull,
...who lives within us all

Richard D. Bach, Author of
Jonathan Livingston Seagull

In memory of Tyler Hill
(www.tylerhill.org, www.clear-cause.org)
and for all people with diabetes
needing help

Table of contents

Foreword

by Viktor Frankl

The fate of people with diabetes is not always easy to bear. They realize this most of all. Therefore, it might happen that they sometimes wonder and ask, "Is it *meaningful* to carry on living with diabetes?" We human beings are the only living creatures searching for the meaning of existence.

Now it has been proven — even on an empirical-scientific basis — that everyone might be able to find meaning in their life and in every situation under any conditions and circumstances, even the most sad and tragic. Our life does not cease to have meaning when we are confronted with an unalterable fate such as an incurable, chronic illness! We cope with chronic illness through faith and behavior adjustments. We can endure it with dignity and courage, or we can simply allow ourselves to succumb to the illness. We bear witness to human capabilities, the possibility to achieve great things through our trials.

A curable illness must be treated; an operable cancer must be removed. We must eliminate the cause and find the remedy for the illness, whenever possible. This is especially valid when it concerns an unalterable fate. *Whenever it is possible to actively shape fate; we must not passively accept it!* We are personally responsible for exceeding the limits, doing all that is feasible, and applying all the advantages of modern medicine. We have the opportunity to transform a personal tragedy into an exemplary triumph!

Our lives are more meaningful when we experience something or someone. In Dr. Howorka's book, the reader experiences the obligation to optimize the metabolic state of diabetes. Whoever adheres to her advice will reap the benefits of their efforts. They will be thankful for her research just as I am thankful to Dr. Howorka. I served as head doctor for 25 years in the same hospital where I had the good fortune of being Dr. Howorka's patient.

Viktor E. Frankl, MD
University Professor, Vienna
1905-1997

Dr. Viktor Frankl was a prisoner of war at Auschwitz. He lost his mother, father, brother and wife. He was stripped of his identity and life's work. Dr. Viktor Frankl is famous for the creation of logotherapy, finding happiness and meaning in any situation, under any circumstance by accepting unalterable fate and working with the future. He died during the same time as Mother Theresa and Princess Diana.

Testimonial

by Helga Grillmayr

350 mg/dl Blood Glucose – Out of Range!
☹ On Conventional Insulin Treatment:
I had feelings of guilt, fear and helplessness. I sometimes waited for hours depressed, without eating, until my blood sugar was "down" again.

☺ Today, with Functional Insulin Treatment:
If it happens at all - immediate, specific correcting to the desired blood sugar level and a quick thought about the possible cause.

Attending a Concert With Friends
☹ On Conventional Insulin Treatment:
I used to quickly inject delayed-acting insulin in the ladies room before the concert. Half an hour later, during the concert and my insulin-to-food interval, I would eat crackers (possibly during the quiet pianissimo), usually 3 carbohydrate choices and 1 fat exchange. You can imagine how much I enjoyed the food or the music. Later, my friends and I would go to an Italian restaurant to eat spaghetti, tortellini or pizza. I would have a bowl of salad with half a slice of bread, a half carbohydrate choice. Oh well.

☺ Today, with Functional Insulin Treatment:
I am no longer obliged to eat an early evening meal. After the concert, I join my friends at the Italian restaurant. I measure my blood glucose level there, inject insulin and eat spaghetti, tortellini or pizza with my friends. It is a great pleasure, both the concert and the meal afterwards!

Appendectomy
☹ On Conventional Insulin Treatment:
I had to rely on my surgeon to understand and control diabetes. I was not confident they could! I worried a lot. I experienced problems fasting before and after the surgery.

☺ Today, with Functional Insulin Treatment:
I would do more frequent blood sugar tests enjoying **no obligation to eat** and experience fewer -- if any -- problems with much less stress. I know what to do.

Spending a Day Hiking With Friends
☹ On Conventional Insulin Treatment:
My friends wait for me to consume my meal in the most impossible places. An hour later, they eat their ham sandwiches at the mountain hut. I nibble on radishes and peppers.

☺ Today, with Functional Insulin Treatment:
I eat what and when my friends eat while keeping my blood glucose close to normal. I simply test and inject insulin to cover the food I eat.

Vacation or Group Trips with Time Zone Changes

☹ On Conventional Insulin Treatment:

On every trip, I had to remember how to adjust my insulin for time of day changes! Did I remember to take along enough food in case of emergencies? My personal "diet" clock ticked endlessly in the back of my head. Poor blood glucose values despite great efforts. Nothing seemed to work! Worries, desperation and resignation resulted. I held feelings of guilt and anxiety about long-term damage.

☺ Travel Today with Functional Insulin Treatment:

I don't worry about time of day changes any more than anyone else. If an emergency happens, I don't have to eat! My blood glucose values are usually in the target range. My HbA_{1c} values are near normal. I live a happy, healthy, normal life without the fear of long-term consequences of high blood sugars. I enjoy traveling!

I could mention numerous uncomfortable situations ***prior to*** my "FIT" era.

Is our diabetes fate unalterable? Do we have to resign ourselves to it? No! Today, there is a form of therapy that frees us from many, though certainly not all, of the problems we face living with diabetes. FIT liberates. It is possible to live freely!

FIT gives people affected by diabetes the opportunity to:

- ☺ Independently control their blood glucose in all situations;
- ☺ Live freely;
- ☺ Be proactive instead of reactive to the effects of insulin;
- ☺ Take personal responsibility for their blood glucose control;
- ☺ Achieve satisfactory blood glucose control with fewer swings between highs and lows and near normal HbA_{1c} results;
- ☺ Delay or prevent long-term complications of diabetes, e.g. blindness, kidney problems and amputations.

There is no dinner for free. No gain without effort. I need to measure my blood glucose level several times a day, inject insulin and above all be sensible. FIT means actively working with your diabetes, easily and effortlessly. It means being personally responsible and independent. Diabetes is no longer unalterable suffering, but a condition under control.

I am personally and infinitely thankful to Dr. Howorka who developed and explained this method to me in 1983. I have held a "normal" metabolic state with FIT since then. My HbA_{1c} values are within or slightly above the reference range for people without diabetes. I have had diabetes for 33 years without long-term complications. I gave birth to a healthy, bright daughter.

On behalf of all of her patients, I thank Dr. Howorka for her dedicated efforts in teaching, advising physicians, diabetes educators and others that help people with diabetes. I also thank her for publishing this book, to help more people manage their diabetes securely and without problems.

Helga Grillmayr, PhD

Deputy Chairwoman of the Austrian Lay Diabetes Society "Aktive Diabetiker"

1. About the first English Edition

In 1993, the Diabetes Control and Complications Trial (DCCT, continued then as EDIC, Epidemiology of Diabetes Interventions and Complications,2005)[1] the largest diabetes mellitus type 1 study in the world was completed in America by the National Institute of Health. This study eventually proved the association between long-term complications upon the degree of glycemic control in insulin-dependent diabetes. Its conclusion: **Good glycemic control is really worth the effort!**

More than 1,400 diabetic patients with good and poor glycemic control participated in the study and both groups were observed for almost a decade. The study revealed:

1. Relatively good control, i.e. an average blood glucose level of approximately 159 mg/dl, can be obtained by most people over long periods of time with multiple daily injections or an insulin pump. EDIC has shown that good glycemic control lessens the probability of macro vascular complications (like heart attack).
2. Improving glycemic control lowers the risks of eye and kidney damage by at least 50%.

The Diabetes Control and Complication Trial, or DCCT and later EDIC was accomplished with great effort and at a great cost. Patients were selected using strict criteria and under close observation. Patients at risk for hypoglycemia were excluded from the study. Many diabetes health professionals say, justifiably, that the results of this study are not easy to translate into practice. However, in this study, intensive therapy without specific education for prevention of hypoglycemia posed an increased probability of serious hypoglycemic episodes.

Our experience with many hundreds of patients using Functional Insulin Treatment mirrored comparable results many years before the DCCT was terminated. However, similar results (with even less hypoglycemia) can only be achieved when a patient is thoroughly trained before any lowering of average daily blood glucose values can be done safely. FIT patients achieve success lowering average daily blood glucose and HbA_{1c} values with less effort and less serious hypoglycemia through comprehensive structured group education and gain a very important advantage – lifestyle freedom to fast or eat whenever and whatever they choose!

A couple of years back, in the UK one of the biggest diabetes trials of the world, UKPDS (United Kingdom Prospective Diabetes Study) has been released, that explored relationships between blood glucose level and blood pressure in type 2 diabetes and its late complications and mortality. In this 20-year trial, 4.000 subjects were randomized into intensified or "usual", less tightly controlled blood pressure and blood glucose therapy groups. The conclusion is similar to the DCCT trial type 1 diabetes results: intensive diabetes therapy (using insulin or sulfonylurea, in obese patients with added metformin) reduces late complications risk by 12 to 25%. Of particular interest is the fact that the more tight therapy caused in patients with elevated blood pressure and type 2 diabetes a further reduction of diabetes-related mortality (by 32%) and a reduction of diabetes late complications by 24-37%. These numbers confirm our long-term experience. Meta-analysis of new studies including even more controversial ACCORD, ADVANCE and VADT confirmed the beneficial

[1] The Diabetes Control and Complications Trial, or DCCT, conducted over ten years by the National Institutes of Health in America proved irrevocably that improved glycemic control postpones diabetes late complications.

effect of good glycemic control, which should be aimed for from the beginning of diabetes.

Our ISO 9001 quality certified International Research Group on Functional Rehabilitation and Group Education documented that an appropriate training for patients with diabetes should last approximately 30 hours. Accordingly, this structured education is only possible in a group approach. Exercises in theory and practice are necessary to learn how to control blood sugar. Our experience shows that conditions in a hospital do not often assist in the learning process or the course itself. The "real life" is outside of the medical center.

This book is designed as an introduction to FIT and a supplementary part of the Functional Insulin Treatment course.

Recent media reports concerning new developments for treating diabetes increased inquiries about FIT training. Another training module called *FIT Update* was specifically developed for FIT course alumni. *FIT Update* is a two-day refresher course and is intended to provide a sort of tune-up for all people with diabetes who are already trained and treated with multiple daily injections or an insulin pump. A book for *FIT Update* needs still to be developed. Please contribute and provide your questions, concerns or innovative ideas on insulin treatment for future FIT book editions, please write!

All courses are offered in German and English to people with diabetes living near or far away or abroad. *FIT-training* is necessary only once but the *FIT Update* weekend is recommended to diabetes patients after an extensive *FIT training* at least once every two years.

I wish you good luck and joy with FIT!

Kinga Howorka, MD, MBA, MPH
Professor of Internal Medicine

1. To the Reader

- ***Do you need to inject insulin?***
- ***Would you like to achieve near normal blood glucose levels most of the time?***
- ***Do you want to avoid long-term complications of diabetes?***
- ***Would you like flexibility and freedom in your life?***
- ***Could you find 5 minutes each and every day for your insulin treatment and glucose self-monitoring?***

If your answer is *yes* to any of these questions, then this book is written for you. You are ready for FIT!

You don't know it, but you have helped me write it, indirectly. You are represented here as a sort of co-author. This book was put together in the question-and-answer format similar to a dialogue. I imagined that we, you and I -- the diabetes patient and the physician -- could discuss the challenges and the problems, the advantages and disadvantages, risks and rewards of self-treatment based on self-monitoring.

You proved to be a fantastic conversation partner! For that, I thank you.

My particular thanks go to the first hundred FIT patients with insulin-dependent diabetes mellitus, who in the early eighties helped us formulate general guidelines for Functional Insulin Treatment. As a result, we offer you a liberating strategy to achieve near-normal glycemia without rigid schedules.

The best method of reading this book is from the beginning to the end, and not skipping about from page to page "diagonally". All new terms are explained at their first mention. A full explanation of new or unfamiliar terminology is provided in the **glossary** at the end of the book. You may assess your knowledge with quick quizzes (**test questions** at the end of chapters 4, 5, and 18. Correct answers are given at the end of the book). If you or your physician need more scientific information about Functional Insulin Treatment, please refer to my professional manual for physicians, ***Functional Insulin Treatment,*** K. Howorka, Springer Publishers, Berlin, English and German editions.

The illustrations and little creatures in this book portray necessary tools for FIT -- a droplet of insulin, a droplet of blood and a little bit of humor. Karl Alex, creator of the droplets, had diabetes, as well. He passed away already. However, his unique pictures will stay with us forever.

All events and people described in the events are fiction. Any similarities between these and real people are purely accidental and unintentional☺.

2. Introduction: What is Functional Use of Insulin?

What do you mean by functional insulin use? What is that?

Insulin is used functionally to imitate insulin production in a healthy person. It is used functionally by *separate* replacement of:

- *Basal* - insulin necessary to sustain the body and keep stable blood glucose while fasting, and
- *Prandial* - or meal-related insulin, and
- *Correctional* insulin, necessary to lower blood glucose levels to the preferred target.

What does that really mean?

You adjust the insulin supply you need with 4 to 6 daily insulin injections or with an insulin pump, which delivers continuously only short acting insulin. The insulin dosage is personalized to your individual requirements based on several daily blood glucose self-measurements, appetite (requirements for food) or fasting needs. One (Lantus®), two (Lantus®, Levemir®) or three (NPH) daily injections of delayed-acting insulin like NPH, Levemir® or Lantus® are needed to cover basal (or fasting) insulin needs. Eating or correcting high blood glucose levels requires rapid-acting insulin such as regular or rapid insulin analog. Corrections are most important: with type 1 diabetes you can expect every second blood glucose measurement to be off target.

Functional Insulin Treatment requires a separate use of correctional, basal and prandial insulin

Functional Insulin Treatment or *FIT* helps you feel good, attain the best possible glycemic control, reduce health risks, and find flexibility and freedom in your life. In the beginning, FIT was defined as functional, "nearly normoglycemic" insulin replacement. Normal glycemia in people without diabetes is between 80 and 120 mg/dl. After eating, their blood glucose can rise to 160-180 mg/dl for a very short time. (Conversion from mg/dl to mmol can be found at the end of this book.) Since these glycemia targets are not applicable for everybody, the name Functional Insulin Treatment describes best the separate basal, prandial and correctional use of insulin for near-normoglycemia and flexibility.

Do I understand you correctly? I must measure my own blood sugar at least four times each day and inject insulin 4-6 times every day or always wear an insulin pump? This seems like a lot of effort...

Not necessarily. Diabetes self-management is much easier today:

- ☺ You can reuse the same syringe with welded-in needle many times until it becomes blunt. Newer thinner needles are much more comfortable. The same applies for insulin pens, of course.
- ☺ Disinfecting the normal "clean" skin is unnecessary. Insulin already contains disinfectants.

- ☺ Only the future supply of insulin must be kept in the refrigerator. Opened insulin cartridges or insulin pens can be carried with you for weeks without problems.
- ☺ You do not have to inject insulin on a rigid time schedule. With FIT you inject when it suits *you*.
- ☺ Modern blood glucose meters are small and give a result in a very short time, many of them are as fast as 5 seconds. Advanced models can store multiple tests in memory for easy recall and averaging. Alternatively, providing the availability of such strips, you can learn to estimate your blood glucose easily without a glucose meter by using specific visual blood glucose test strips and comparing the result of the test strip to a color scale; more details at e.g. www.betachek.com.

That said, you will need to accept some "uncomfortable seconds" every day due to more frequent shots and self-testing your blood glucose. For most people, gaining lifestyle flexibility and near normal blood sugars with FIT is worth this minimum discomfort. Many people agree that it is much easier to inject when you eat and to test more often than it is to follow the demands of a rigid insulin and diet schedule. But with FIT you need to learn to make the right choices on insulin dosages in your everyday life.

What advantages do I get in comparison to my current 2-4 shots per day with conventional treatment?

You will have near normal blood glucose levels much more often and a much more flexible way of life. You can sleep in. You can eat what you want, when you want or not eat at all! You can live your life much like people without diabetes. Are you prepared to invest 3-5 minutes of self-treatment per day and attend a FIT education course? And, long-term complications of diabetes are minimized or avoided altogether.

Of course! I can spare 5 minutes each day, I waste more time trying to eat every three hours on schedule. If FIT is so easy and effective, why didn't someone do this before?

In the last two decades it became clear how easy it really is to live with "unstable diabetes", using insulin injections or insulin pump, and blood glucose monitoring. No ice box, disinfection or always new needles are necessary. The introduction of self-monitoring blood glucose in the late 70's of the last century was extremely important. Even now, with continuous glucose monitoring it became eventually clear that with "unstable diabetes" the routine blood glucose correction – when hyperglycemic – is essential. By the way, that was proposed by us in early eighties, some 25 years ago already...

Functional Insulin Treatment can be achieved both, with injections or with an insulin pump, and with blood glucose self-monitoring

People with diabetes began to personally contribute to their glycemic control. Insulin pumps are becoming smaller, more convenient and increasingly more comfortable. Insulin pumps come very close to simulating insulin production in a healthy pancreas. Thus, experience with pumps has been important in the development of FIT. Later on, injecting insulin became much easier. It became clear that multiple daily injections with insulin pens is often a preferred method to carry out FIT. In any event, insulin pumps and pens are only instruments. You must have comprehensive diabetes education to make accurate informed and effective decisions.

Our first patients clearly indicated their desire for a more flexible lifestyle. The chance to eat whatever and whenever they wanted, or to fast, was more important to them than the small inconvenience of frequent insulin injections or blood self-monitoring. FIT patients successfully learn to maintain excellent pre- and post-meal blood glucose values and to correct swings immediately. FIT patients have more confidence in their ability to prevent short-term treatment complications like keto-acidosis, severe hypoglycemia and long-term complications of diabetes like blindness, kidney failure, amputations, or premature death. FIT diabetes patients live more normal, healthier lives in all situations!

What particular discoveries led to the development of FIT?

Investigation of insulin produced by healthy persons established a model of functional insulin therapy for diabetes. The catalyst for FIT was a clear understanding of the amount of insulin produced while fasting and eating. Every person with diabetes needs to understand the function of a healthy insulin-producing pancreas. It is necessary to have this understanding to imitate with the least possible effort this "automatic" production of insulin production and regulation.

Decades ago, Dr. Stolte, a pediatrician from Breslau, Poland tried to improve diabetes treatment with independent insulin dosages and self-monitoring. Champions of such modern therapy were often misunderstood. The necessity for feedback between glycemic self-monitoring and insulin dosage went unrecognized. The term "free diet" obtained notoriety and was associated with poor glycemic control from these early attempts at flexible insulin therapy.

Coming back to the facts: FIT has been developed by me in Vienna 1982-83 under the best “sources” for diabetes research and education: Jay Skyler, Miami (algorithms for insulin dosage), Werner Waldhäusl, Vienna (data on insulin production of healthy men), Richard Bernstein Mamaroneck, USA (correctional use of insulin) and published in 1984. My books for patients (Kirchheim publishers, Mainz) and for physicians (Springer publishers, Berlin) appeared first edition in 1987.

Isn’t FIT the same like the intensified insulin therapy?

FIT means functionally separated use of insulin either for blood glucose correction or for basis or for eating. The definition of "intensive insulin therapy" is frequently misunderstood and varies depending on the interpretation of the teacher or doctor and the country. There are some historic reasons for that. Indeed, the introduction of intensive insulin therapy by Jay S Skyler in 1981 had a significant impact on development of FIT. Later on, in eighties and nineties of the last century, the originally “intensified” therapy has been systematically adapted by some of FIT advantages. So the transitions between these two therapy strategies became more continuous.

Interestingly, decades later, in some European countries at least one national "inventor" and self-declared original author of FIT therapy could be found. In fact, it is a very satisfying and enjoyable feeling to see, how well the idea of functional insulin therapy has been accepted and established☺. Or were that just parallel developments?

Rules for self-adjustment of intensive treatment has been defined by Jay Skyler as follows:

APPENDIX A

Algorithms for Adjusting Insulin Doses under Various Treatment Regimens

Regimen 1. Split-and-Mixed Regular and NPH or Lente Insulin

INTENSIVE INSULIN THERAPY

Target Blood Glucose Levels	
Before breakfast	70–110 mg/dl
Before lunch, supper, and bedtime snack	70–130 mg/dl
One hour after meals	≤ 180 mg/dl
Two hours after meals	≤ 150 mg/dl
2 AM to 4 AM	> 70 mg/dl

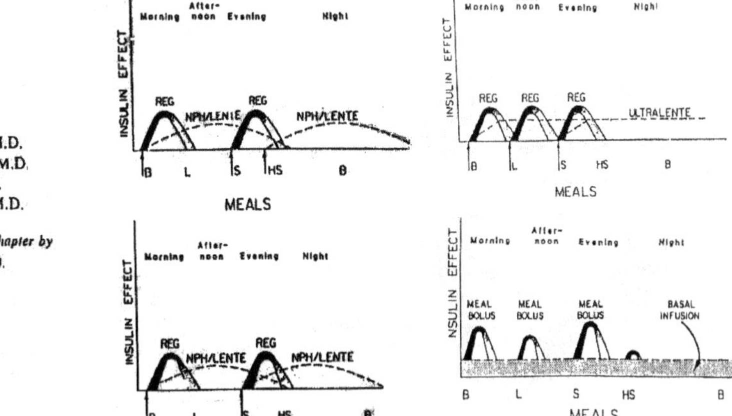

David S. Schade, M.D.
Julio V. Santiago, M.D.
Jay S. Skyler, M.D.
Robert A. Rizza, M.D.

with an introductory chapter by
Paul Haycock, M.D.

1983
Excerpta Medica

Fig 2.1. Algorithms for adjusting insulin doses with intensive insulin therapy have been defined by Jay Skyler. Neither flexibility in eating schedule nor blood glucose correction were accepted treatment goals. Adapted from: Schade DS, Santiago JV, Skyler JS, Rizza RA: Intensive insulin therapy. Excerpta Medica, 1983

With these schedules of multiple daily injections of rapid- and delayed-acting insulin, is it necessary to eat regularly?

Yes. In the eighties, the times of IIT developments, neither correction of hyperglycemia nor variability of eating schedule have been defined as therapy goals. Treatment schemes based on use of Ultralente-type insulin (delayed-acting insulin; no longer available today) were critical for further FIT developments.

Intensive therapy is covered in more detail later on in Chapter 4 on Basic Diabetes Education. By the way, the term "intensified" is rather misleading when it is applied to FIT because the *total effort* devoted to Functional Insulin Treatment by the patient actually *decreases*. FIT patients are not required to adhere to strict and sometimes oppressive schedules and diets for their insulin therapy. In summary, for functionally separated use of insulin for fasting, eating and immediate correction of high blood glucose with the aim of near normal glycemia and life style flexibility, we propose the term: Functional Insulin Treatment - FIT.

Do I need specific education or training for FIT?

Yes. FIT is an advanced form of therapy for people with diabetes. It is recommended for participants who already perform blood glucose self-monitoring several times a day and self-adjust insulin dosage accordingly. It is very advantageous if you have already taken part in a proper "basic" diabetes group education. Quick quizzes at the end of Chapter 4, Basic Diabetes Training, assess your general knowledge of diabetes and self-care protocols. Qualified educators teach FIT in a group setting. Phase 1 and 2 of FIT training includes approximately 28 hrs of structured education.

Can I learn and apply FIT from this book only?

Possibly. But we do not recommend it! This book was written to introduce you to FIT training, with a special emphasis for those who are treated in clinical settings. It is *not* a self-training tutorial. We recommend medical supervision and physician involvement. If you are fortunate enough to have a diabetes team with a dietitian, diabetes educator, nurse and endocrinologist helping you, this is superb! Several blood glucose experiments, so called 'insulin games' are useful to learn FIT quicker. This all is very similar to getting your driver's license...or have you done it only by following a manual without any practical exercises? You will be more successful with your diabetes team's support and active involvement. Teaching independent, functional use of insulin is relatively new. Misunderstandings can develop with any innovation. This book is designed to avoid such misunderstandings and to introduce you, your family members and anyone else interested in diabetes care to this liberating and effective form of personalized insulin treatment.

3. Overview of the Functional Insulin Treatment Program

I am familiar with carbohydrate counting in diabetes diets. What do I have to learn to attain near normal blood glucose levels and a more flexible, normal life?

We have created, along with our patients, a **FIT Training Program** to enable you to learn functional use of insulin, contingent on current blood glucose level, with the most possible fun and the least possible effort. The important features of the training are:

1. An outpatient group education with approximately 8-15 people per group.
2. Numerous practical exercises. At least one-third of the course is devoted to exercises, insulin games and practice-oriented methods including daily practical discussion of individual case histories.

Learning to drive also takes time. Without practical exercises no one can maneuver safely. FIT training is comparable. The program consists of two outpatient phases. **Phase One, Basic Diabetes Education** lasts two days, approximately 12 classroom hours. It is an essential and comprehensive preparation for the switch to functional insulin use. It includes the following contents:

1. An overview of the production of insulin and insulin action in a healthy person;
2. An overview of the different types of diabetes, their heredity and how diabetes develops;
3. Self-monitoring: blood glucose, urinary glucose and ketone testing for effective diabetes control;
4. Consequences and symptoms of insulin deficiency; DKA (i.e. diabetic keto-acidosis), and hyperosmolar coma. Hypoglycemia and how to avoid it.
5. Nutrition and meal planning as defined for conventional treatment of insulin dependent diabetes (carbohydrate counting and basics of caloric content of food);
6. Current possibilities and strategies of insulin treatments
7. Associated diseases (high blood pressure, blood lipids, metabolic syndrome, new treatment targets) and their interrelationship with diabetes.
8. Late complications of diabetes and how to avoid them.

By the third day, you should have sufficient information on current treatment alternatives for your diabetes. **You** can now choose the form of insulin therapy most desirable to **you**. Most people opt to self-regulate their insulin dose to enjoy a flexible diet with lots of variety and directly influence their blood glucose levels. Thus, they opt for FIT. In this case, frequent injections (4-6 daily), or pump, frequent blood glucose tests (4-6 daily) and the responsibility of taking charge of your diabetes need to be accepted. The only alternative widely used in the US, acceptable in the context of insulin-dependent diabetes, the conventional-intensified insulin therapy, which recommends 2-4 injections per day and a minimum of 4 blood tests daily, is comparable to the standard as far as the daily time

FIT-education in functional use of insulin: either for fasting, corrections or for eating

and efforts expenditure is concerned. However, with intensified therapy, much higher consistency in food intake is required. See Fig 2.1

And the second phase?

Phase Two of the program is the real **FIT Training**. You will learn how to **functionally** use insulin to cover your body's needs for fasting, eating or correcting high blood glucose levels. To be FIT you must answer some important questions to make appropriate choices for insulin doses. These questions are in the following table:

1. BASAL INSULIN (for fasting)
How much insulin — and which type — do I need to inject even if I don't eat?
The basal insulin can be covered with delayed-acting insulin; a small amount of rapid-acting insulin in the morning is often required in addition to this.

The action profiles of delayed-acting insulin (NPH and analogs Lantus® and Levemir®) on blood glucose need to be known.

2. PRANDIAL (meal related) INSULIN
How many units of rapid-acting insulin do I need for 1 carbohydrate choice?

It is also advantageous to know:
Which insulin (regular or a rapid-acting insulin analog)?
Which time-interval, if any, between injecting and eating; how to inject?
What influences the absorption of my insulin?
Carbohydrate counting necessary;
In low-carbohydrate meals the amount of non-carbohydrates should be also considered; how?
Why it is not good to eat when I have high blood sugar?

3. CORRECTIONAL INSULIN *(for controlling current blood glucose level)*
What is my target for blood glucose corrections (before meal, after meal)?
What is the effect of 1 unit of rapid-acting insulin on my blood glucose?
What is the effect of 1 carbohydrate choice on my blood glucose?

For assessing efficiency of blood glucose corrections:
What is my target range for daily mean blood glucose (MBG)?

Fig.3.1. Begin FIT only when you and your physician or CDE (certified diabetes educator) know the answers to these questions.

I got it! If I need two insulin units for 1 carbohydrate choice, then I will need four insulin units for 2 carbohydrate choices. Is that right?

Exactly! This is necessary to cover your insulin need for food. On the other hand, you must know the effect that one unit of rapid acting insulin has on **you,** how high **your** current blood glucose level is and **your** recent blood sugar correction target to effectively manage your diabetes and be FIT. You can then calculate *exactly* how many units of insulin **you** need to lower **your** blood sugar level at that point in time. The duration of your rapid-acting insulin (regular or rapid-acting analog) must always be considered before injecting "correcting" insulin!

Every participant of the FIT education program receives his/hers personal rules for insulin dosage. Everyone should know when, how much and what type of insulin is needed when fasting, eating or correcting occasional unacceptable high blood sugars. These personal rules of insulin dosage are called "algorithms for FIT". They describe how to cover your basal (fasting), prandial (food related) or correctional insulin needs.

If I had my "algorithms" for insulin dosage could I begin immediately?

It's not quite that easy. Your algorithms change with different circumstances in your life. You are really FIT only when you can correct your unacceptable blood sugars and *adjust your algorithms* when your insulin requirements change. The second FIT training phase uses a few insulin games to teach you how to make accurate judgments of your algorithms to achieve good glycemic control, despite of changed needs.

There are simple equations to derive your basal, prandial and correction FIT algorithms from your total daily insulin need. You will find them at the end of the book. Your physician will help you with estimation of your initial FIT algorithms at the end of the Basic Diabetes Education.

In our experience, about 3-4 days including at least 20 hours of teaching, are necessary to learn the functional use of insulin. It will require about 30-35 hours of structured education if you include twelve hours of Phase One Basics. The second phase, i.e. the FIT training is extremely important!

How do I begin?

Insulin is a very effective medicine. Do not underestimate it! To reach your aim as quickly as possible, look for a health professional to help you. You will not get rid of your diabetes despite FIT, you will only learn how to control it. For more information about how you can find a FIT advisor to help you, preferably in a training center or a diabetologist office, please check also our website *www.diabetesFIT.org*.

Strategies of Insulin Treatment
Conventional
Functional

2. Phase One - Basic Diabetes Training

Must I go through Phase One of the FIT Program, the Basic Diabetes Training? I have already attended a diabetes training course and think I know enough already.

Might be. However, there is a quick quiz at the end of this chapter to help you assess your readiness to begin Functional Insulin Treatment or FIT. If you do not excel at this quiz (e.g. you reach less than 95% of correct answers, the test is rather easy), I advise you to study this chapter again very carefully (especially the part about insulin types and food), attend a basic diabetes training seminar and read one of the books referenced at the end of this chapter. Perhaps we should start with a short summary of diabetes. Do you have any questions?

What is insulin?

Insulin is a hormone produced in the beta cells of the "Langerhans" islands in the pancreas and is released directly into the bloodstream.

How does insulin work?

Insulin transports glucose, or energy, from the blood into the cells, lowering blood glucose. Almost all tissues in the human body require insulin.

Why can't it be swallowed?

Insulin is a protein. Your stomach would digest it if it were swallowed making it useless.

Does a healthy pancreas always produce the same amount of insulin?

No. People without diabetes produce a limited and somewhat constant amount of insulin all the time (i.e. between meals and during sleep as well as for a greater amount of food intake). After eating, carbohydrates are transferred from the intestines to the bloodstream; the increase in blood glucose causes an immediate increased production of insulin depending on the content of the meal. The balance of glucose under normal circumstances is maintained by a sensitive closed control loop, where an increase in blood glucose leads to the release of insulin. A decrease in blood glucose, on the other hand, elicits inhibiting insulin production and the release of counter-regulatory hormones, the most important hormones being glucagon and adrenaline.

What factors influence the production of insulin in a person with a healthy pancreas?

The amount of insulin produced depends, normally, on the amount of food supplied, mainly the amount of carbohydrates. Approximately half of the daily insulin discharged from the pancreas is produced after eating to meet meal-related insulin needs. This is referred to as *prandial* insulin. The other half corresponds to the fasting insulin requirement, referred to as *basal* insulin. Other factors influencing insulin requirements are exercise, stress or illness, and alcohol consumption.

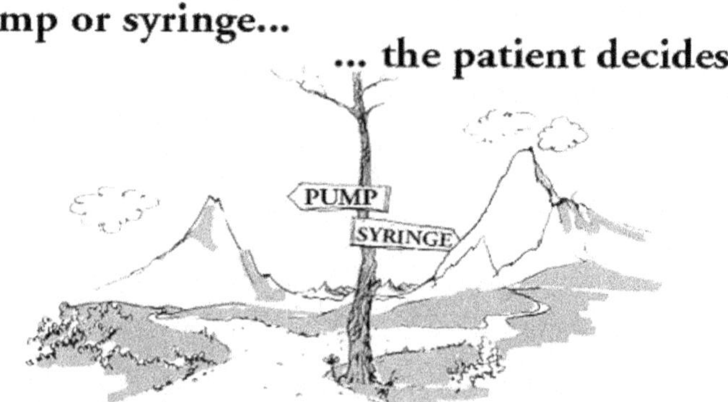

Daily profile of insulin and glucose in a healthy person

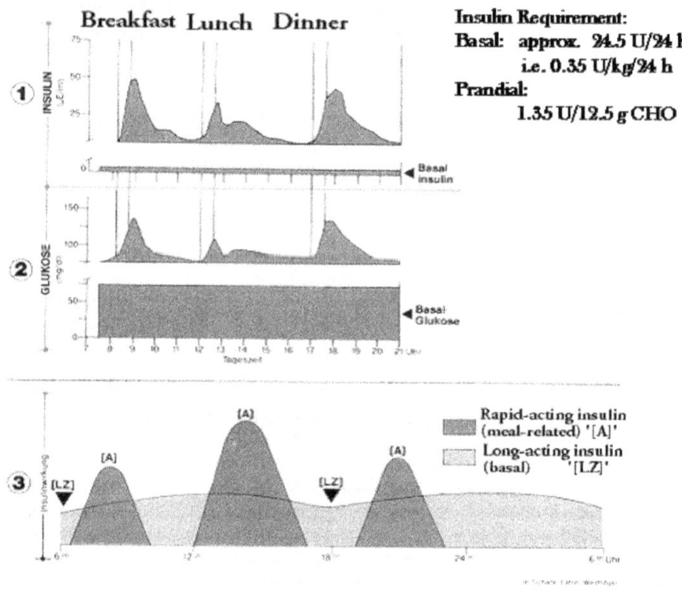

Fig. 4.1. The production of insulin (1) and glucose (2) in the blood in a person without diabetes. Modified after Waldhäusl et al, 1979.

What happens when someone without diabetes drinks alcohol?

Alcohol inhibits the production of glucose in the liver. Less glucose in the blood requires less insulin. Insulin needs decrease when alcohol is consumed: The liver acts as a depot where carbohydrate supplies are stored in the form of reserve glycogen. From this "depot", a certain amount of glucose is released into the blood to meet the energy needs of cells in the human body. Insulin transports glucose into the cells. Insulin enables glucose to be converted into energy or fuel by the cells.

Does alcohol reduce insulin requirements even for people with diabetes?

Yes. People treated with insulin need to either reduce their dosage of insulin or eat more carbohydrates than usual if they want to drink alcohol. It is advisable not to drink more than two drinks and exercise caution when consuming alcoholic beverages.

Alcohol lowers blood glucose

What happens when you exercise more than usual?

Exercise increases the effect of insulin. Small amounts of insulin are needed to increase the transport of glucose into the cells during physical activity. In a healthy pancreas, insulin production decreases during exercise so there is less concentration of insulin in the blood.

Will someone treated with insulin develop hypoglycemia if he or she does not eat more carbohydrates before exercise?

Correct. Alternatively, they could reduce the dose of insulin appropriately.

Do all people with diabetes produce insufficient insulin?

They either produce too little insulin or the insulin is ineffective in regulating cellular absorption of glucose. People with type 1 diabetes have high blood glucose levels due to an absolute insulin deficiency. In type 2 diabetes, high blood glucose levels are due to a decreased insulin effect or insulin resistance.

How do I know if I have type 1 or type 2 diabetes?

It is not always easy to differentiate between the two. Type 1 diabetes occurs much less frequently than type 2. Less than 10% of all people with diabetes are afflicted with type 1. Type 1 diabetes occurs suddenly and most often during adolescence. In the past, it was known as "juvenile-onset" or "insulin-dependent" diabetes. People affected by type 1 diabetes are usually thin. Acetone (=ketones) appears in the urine, indicating insulin deficiency and thus burning of fat instead of glucose. Glucose is released into the urine when certain blood glucose levels, or "kidney thresholds" for glucose, are exceeded. Glucose discharges into the urine increasing the volume of urine and often dehydrating the person. To differentiate between the two diabetes types you can ask for measurement of C-peptide in your blood. Even if you inject insulin (which cannot be differentiated from your own) with type 1 diabetes your C-peptide level will be low or not detectable. C-peptide is produced in your pancreas from proinsulin (from each molecule of proinsulin one molecule C-peptide and one molecule insulin will be delivered). In type 2 diabetes with typical insulin resistance, a high level of C-peptide indicating high insulin level will usually be assessed.

Type 1 diabetes (insulin-dependent diabetes) is an auto-immune disease

Is the need to urinate often associated with a discharge of glucose in the urine?

Yes. Most people notice excessive thirst before they recognize the large volume of urine or urge to urinate often.

Why do people with type 1 diabetes lose weight before the diagnosis is made?

The lack of insulin causes a "state of hunger" in the tissues and cells. Excessive glucose is found in the blood, but it cannot be transported into the cells without insulin. The cells are "starving". It is impossible to live without insulin; the available energy supplies in the body are depleted. These energy supplies are stored in nearly all cells in the form of fat. Acetone, or ketones, in the urine indicate that the body is obtaining most of its energy from fat instead of glucose. Glucose in the urine resulting from excessive high blood glucose also has a diuretic effect. Weight loss is partially caused by the loss of water and using fat for energy.

Is acetone (ketones) in the urine always a sign of too little insulin?

Generally, any person can discharge ketones in the urine when food intake is inadequate. If diabetes is detected and the nutrient intake is sufficient, ketones discharged in the urine are often a sign of absolute insulin deficiency, indicating the existence of type 1 diabetes. Coexistent signs are hyperglycemia, high blood glucose, and glycosuria, or glucose in the urine.

There are two types of diabetes...
THER

What are the characteristics of type 2 diabetes?

Type 2 diabetes, known in the past as "adult-onset" or "non-insulin-dependent" diabetes, occurs primarily after the age of 40. People with this type of diabetes are often overweight, at least at diagnosis. Due to excess weight and a certain genetic predisposition, insulin loses its effectiveness, which results in the development of a "relative" lack of insulin and a subsequent increase in blood glucose. In contrast to people with type 1 diabetes, those with type 2 diabetes are not generally treated by insulin from diagnosis but may be prescribed insulin later, if necessary. Here, insulin is only necessary when metabolic control is unsatisfactory after dieting, significant weight loss, or treatment with tablets (oral agents). Type 2 diabetes does not begin as abruptly as type 1 diabetes. Here symptoms develop slowly. Increased blood glucose values are often detected by chance.

Does type 1 diabetes always need to be treated with insulin?

At the moment, the answer is "yes". This form of diabetes develops when more than 80% of the insulin-producing beta cells in the pancreas are destroyed.

Ketones in urine show insulin deficit! Each type 1 diabetes patient needs strips for urinary ketones assessment

How are the cells destroyed?

It is generally accepted that type 1 diabetes is an "auto-immune" illness, i.e. the body's defense or immune system is mistakenly mobilized against its own beta cells, before symptoms even appear. It is not yet known what factors cause this "self-destruction" of insulin producing beta cells.

Can anyone prevent this mistaken defense and self-destruction in time?

In theory, yes. This process could be suppressed with an "immune intervention" - an ongoing, long-term, medicinally controlled change of the self-defense mechanism. As I mentioned previously, the symptoms of type 1 diabetes only develop after more than 80% of the available beta cells have been destroyed. At this point, immune intervention is too late even if it is started without delay. Currently, studies with an immune intervention component starting as early as possible are being conducted in diabetes prevention trials. In a few years, more will be known about the value of this type of intervention.

Type 2 diabetes implicates *always* the presence of metabolic syndrome with elevated blood pressure and blood lipids

So, people with type 1 diabetes need insulin to live. What does the future hold for these people?

The quality of life and life expectancy of people with diabetes are dependent, largely, on the onset of diabetes-related complications, which result from a chronic increase in blood glucose over time. Little is known about other influences such as hormonal and genetic factors. Disorders of the capillaries or small blood vessels, also known as "microangiopathy", can cause late complications affecting the eyes, nerves, feet and heart. Changes in the large arteries (atherosclerosis) occur more often, earlier and more severely in people with diabetes. Damage to nerves or diabetic neuropathy, affects the sensitive and autonomic nerves which support the internal organs of the body. Good metabolic management in the early stages of diabetes is crucial to preventing irreversible late complications and organ damage.

Hint for central obesity is waist circumference in Europe in women >80cm in men >94cm (in the US 88 and 102 respectively ☺)

If late complications already exist, what is the point of maintaining good blood glucose control?

The progress of late complications can be slowed down, or avoided altogether by tight control, particularly when the vessel changes are not advanced and/or late complications are treated early enough (e.g. laser or light beam coagulation of the retina, reduction in blood pressure, etc.). Avoiding smoking is another way to delay or prevent complications from occurring. And keep your blood pressure low. Even slightly increased blood pressure enhances microvascular diabetes complications.

When do late complications of diabetes generally appear?

This depends mostly on individual glycemic control. Retinopathy or microangiopathy (in the retina, for example, visible with an ophthalmoscope) can frequently occur after 7-15 years of conventional diabetes treatment and sub-optimal glucose control.

Does good glycemic control offer complete protection from vessel damage?

There is strong evidence that well-controlled diabetes provides protection from late complications even decades after the onset of the illness. Of course, this question is difficult to answer since achieving really good glycemic control has only been possible in recent years. Measuring metabolic control in people with diabetes over longer time periods is now objectively possible by determining the glycated hemoglobin (HbA_{1c}) level. The Diabetes Control and Complications Trial or DCCT (mentioned earlier), clearly showed that the damage to the retina or kidneys can be reduced greatly by excellent metabolic control, which is associated with lower levels of HbA_{1c}.

The same is valid for type 2 diabetes (UKPDS, United Kingdom Prospective Diabetes Study 1998) and the outcomes can be even improved with blood pressure control. The follow-up study of DCCT, EDIC (Epidemiology of Diabetes Interventions and Complications, 2005) has shown that blood lipids are also relevant. All that means: the blood pressure and blood lipids have to be normalized if you wish to stay healthy. For that purpose we have developed new educational modules: the 'hyperlipidemia' and 'hypertension' group education modules.

Meta-analysis of new studies including even more controversial ACCORD, ADVANCE, VADT confirmed the beneficial effect of good glycemic control, which should be aimed for from the beginning of diabetes.

Based on recent scientific studies about the development of late complications in diabetes, what conclusions can be drawn for blood glucose control?

First, because the insulin requirement in type 1 diabetes depends on many factors and fluctuates, it is necessary to balance the day-to-day fluctuating insulin requirements with appropriate blood glucose corrections. Second, because of these fluctuations in blood glucose and insulin requirements, treatments with a fixed schedule for meals and injections usually do not solve the problem. New types of treatment offer flexibility and a good quality of life together with optimal metabolic control. And third: blood pressure and blood lipid control are essential as well.

Which types of treatment do you mean exactly?

The best metabolic results are achieved by a "tripartite", or threefold approach. The person using this approach injects (1) "basal" insulin for fasting, (2) "prandial" insulin for eating, and (3) "correctional" insulin for counteracting hyperglycemia (an excess of glucose in the blood). Only when you separate insulin (the basal from the prandial and from the correctional) can you use insulin effectively to regulate your blood glucose and use either carbohydrates or rapid-acting insulin to bring your blood glucose level into the desired target range. This is currently the only way to achieve tight blood glucose control your blood glucose possible in a normal to near normal fashion.

Is replacing fasting insulin, prandial insulin, and correctional insulin by the diabetes patient called FIT?

Yes. People with diabetes using this therapy are self-reliant. They have the knowledge and skills to make a direct, immediate and accurate impact on their blood glucose. They have the advantage of personally monitoring and controlling their blood glucose every day while living a healthier, more flexible and independent life.

How do they "control" their blood glucose?

To control or "direct" something, you need to know where you are and where you are headed. Blood glucose monitoring in FIT requires a certain "position determination." This means measuring your blood glucose to find out where you are at the moment. Knowing your present blood glucose target area is necessary to determine whether you need to correct. An insulin deficit, or difference between high blood glucose and the desired target, can be corrected by using the following rule: "One unit of rapid-acting insulin lowers my blood glucose by ____mg/dl."

How do I evaluate my glycemic or metabolic situation?

You test the following:
(1) Blood glucose
(2) Urinary glucose
(3) Urinary acetone or ketones.

For details, please refer to *further reading* at the end of this chapter.

Does blood glucose self-monitoring offer any advantages over –historical– urinary glucose testing? Finger pricks are a little uncomfortable.

Test results from glucose in the blood or urine do not provide the same information. The type of metabolic self-monitoring that is best for you depends mostly on the selected strategy of the treatment and what you want to accomplish. A blood glucose test provides a more accurate and current indication of your blood glucose. A positive urine glucose test tells you that your blood glucose exceeded your "kidney threshold for glucose" since you last urinated. Your kidney threshold is the blood glucose level where your kidneys begin flushing glucose from the blood into the urine. Urine glucose testing is not so useful as blood glucose testing as the result indicates that blood glucose was elevated up to several hours before. Blood glucose testing indicates blood glucose levels in the last few minutes, just now.

What is the "kidney threshold" for glucose?

The "kidney threshold" is different for everyone. It is generally between 170 and 220 mg/dl. There are people with diabetes who show an unusually "low" (e.g. below 150mg/dl) or unusually "high" (above 250 mg/dl) kidney threshold. Blood glucose self-monitoring is preferred to avoid interpretation problems. Testing urine for glucose is meaningful to determine if you experienced high blood glucose exceeding your kidney threshold since your last urination without a finger prick. It tells you that a "correcting" insulin dose may be needed. But if you choose a treatment form with variable food intake, urine glucose testing alone is not satisfactory.

What are the main advantages of blood glucose tests?

Obtaining accurate current blood glucose values. You can accurately and immediately compare it to your target value and react appropriately. For example, you can correct high blood glucose values with rapid-acting insulin if necessary or correct low blood glucose values with carbohydrates. Blood glucose tests become indispensable the more you strive to achieve both normoglycemia (near normal blood glucose) and flexibility of lifestyle.

If only I did not have to prick myself...

There are many ways to make blood glucose self-testing easier, more practical and less painful in everyday use:

1. Do not use lancets alone. Use a lancet device specifically developed for blood glucose self-monitoring.
2. You can use a fine pricking device numerous times to prick your finger until the needle becomes blunt.
3. Disinfecting the skin before pricking the finger is unnecessary if your hands are clean.
4. Use electronic blood glucose meters to provide easy, quick and accurate results.
5. For blood glucose tests, use a different finger each time. Many people find it less painful to prick the side of their fingertips.
6. If available, learn to estimate your blood glucose levels with a visual and color scale (see www.betachek.com). This is important when you lose your meter. Test your ability to visually compare your blood glucose on a manual strip with your physician's value. Visual blood tests require a watch with a second hand.

When and how often do I need to test my blood glucose?

That depends on these factors:

1. Your chosen glycemic target, and
2. The treatment strategy applied.

There is some consensus, that on intensified conventional therapy, a minimum of two or three insulin injections are needed daily if there is an absolute lack of insulin. The "minimal" treatment consists of rapid- and long-acting insulin, each twice daily.. At least four glucose tests are essential each day including fasting (ie upon waking and before eating), before lunch, before dinner and later in the evening. This conventional form of therapy includes a fixed diet plan and schedule to match the peak of foods consumed to the peak of the insulin. This type of treatment is described by Skyler algorithms (Fig.2.1)

On the other hand, FIT enables daily flexibility with food and requires varying doses of insulin via multiple daily injections or an insulin pump in a more physiologically correct way. At least 4-5 daily blood glucose tests (better: around 6 daily tests) are necessary for excellent long-term glycemic control.

When is the most important time for me to test my blood glucose on FIT?

With blood glucose dependent insulin dosing, the most important blood glucose check-up is late at night. This test helps you make an accurate blood glucose correction if blood sugar is outside your target and thus to obtain good blood glucose values while you sleep. Sleeping equates to one-third of your lifetime! Assuming that the correct basal insulin doses are applied, you can easily bring your blood glucose into your target range during the night and rest assured that hypoglycemia will not occur.

The second most important blood glucose measurement is in the morning. This test is often termed the fasting test. Fasting blood glucose values vary the most. In addition, two to three tests spread throughout the day are necessary. It is advantageous, but by no means necessary, to measure your blood glucose before every meal. Approximately one-third of all blood glucose measurements need to be taken randomly after eating to estimate whether the fast acting insulin was correct for the rate of nutrient absorption in a particular meal and later on, for the chosen amount of carbohydrates.

My blood glucose fluctuates a lot. Is this harmful?

There is no evidence, at present, that blood glucose fluctuations per se are harmful if you correct them immediately. There is evidence that persistent hyperglycemia (high average blood glucose and glycosylation of the tissues) will lead to late complications. You can avoid such damage by achieving near normal long-term control of your blood glucose, and hence, near normal HbA_{1c} levels.

What is an HbA_{1c}?

HbA_{1c} is the abbreviation for glycated hemoglobin A_{1c}. It was previously referred to less accurately as HbA_1. The HbA_{1c} value gives information about the average blood glucose level within the last six to eight weeks before blood sampling. This corresponds to the lifetime of red blood cells. HbA_{1c} is a blood test used to measure how tightly your diabetes is controlled. It measures how much protein in your red blood cells is eventually binding with glucose (="glycosylation"). This measure, to a certain extent, allows estimation of average blood glucose levels. People without diabetes have average blood glucose of 85 mg/dl, about 5% of all hemoglobin molecules are irreversibly bound to glucose; hence, they are called "glycated". The glycosylation corresponds to the "sugaring". Well-controlled diabetes can achieve almost normal blood glucose. In this case, the amount of glycosylated hemoglobin or HbA_{1c} amounts to approximately 5-7%. More glucose binds to the hemoglobin when the average blood glucose is 30-60 mg/dl higher than normal. Then, the HbA_{1c} value is increased. Each laboratory defines its own "normal" ranges for HbA_{1c} values for their methods, determined from tests of people without diabetes. This is referred to as the "reference range for HbA_{1c}".

How do I know how to interpret my HbA_{1c} result?

In people with diabetes, a normal HbA_{1c} value is characteristic of "excellent" metabolic control, although this might be related to the danger of hypoglycemia, or low blood sugar. Metabolic control within 1% of normal HbA_{1c} ranges might be considered as "very good" to "acceptable". Measures higher than 1% are generally considered as "satisfactory" control. And, values higher than 2% above the upper normal Hba_{1c} range indicate "unsatisfactory" or "poor" metabolic control. In general, there is a strong correlation between your average blood glucose and HbA_{1c}. Recently, the term estimated average glucose eAG has been developed (see table in Appendix). With eAG you can "translate" your HbA_{1c} to mean (average) blood glucose.

Are HbA_{1c} values only determined in the laboratory?

At the moment, yes. The determination is quite complex, although most laboratories are capable of recording this parameter. Recently, a simple finger prick has been shown to be an acceptable alternative to venous draws. The results may be than available immediately in some cases.

How often do people with type 1 diabetes need to determine their HbA_1c values?

At least once every three months, a total of four times per year. Regular HbA_{1c} tests enable self-treatment to be assessed for accuracy. Single peak values of self-blood tests are often overestimated. You should need to achieve a low, nearly-normal, HbA_{1c} level but without experiencing hypoglycemia. HbA_{1c} values within the normal reference range can often be associated with over-insulinization and repeated episodes of hypoglycemia. In our experience, it is sensible and safe to strive for HbA_{1c} values within 1-1.5% of the upper range for normal. If you compare this to the estimated AG, you will receive the number of around 160-170 mg/dl for mean blood glucose as still acceptable.

It is too bad that HbA_{1c} tests can only be checked in the laboratory.

If you monitor your blood glucose values several times a day, you can calculate your average blood glucose values every day from your measurements. This average daily blood glucose value allows you to make independent decisions, changes and conclusions regarding your rules for insulin dosage.

What should be my target for average daily blood glucose?

The target **mean blood glucose**, MBG, is dependent on the strategy of insulin treatment. We recommend daily MBG no lower than 100 mg/dl and no higher than 160(170) mg/dl for people educated in FIT, using multiple daily injections or an insulin pump. Exceptions apply to pregnant women and to people who experience hypoglycemic unawareness and severe insulin reactions with unconsciousness. From our experience, self-monitored MBG values, including post-meal values, at approximately 130-160 mg/dl are the safest for the majority of patients with type 1 diabetes. You will obtain a meaningful parameter of your blood glucose control when you average MBG values over the last 7 days and achieve a **MBG value for the week**. Control of your diabetes with FIT is considered good if this value is between 110 and 160 (170) mg/dl. This refers to blood sugar self-measurements, of which approximately one-third are taken after eating. Average blood glucose values are obviously lower when you measure your blood glucose only before eating. The MBG is generally higher with conventional therapy using two to three shots per day because of the increased risk of hyperglycemia. With FIT we found a high correlation between self-monitored/calculated MBG values and HbA_{1c} (r=0.6).

By the way, would you believe that the greatest study on type 1 diabetes, DCCT, only achieved an MBG of 159 mg/dl and a HbA1c of 7,2% in the 'intensified' treatment group? This is what our FIT patients achieve routinely...

It looks like I need to buy a calculator.

That's right. You obtain important, meaningful results about your overall control by calculating your daily and weekly MBG. The MBG provides you with a reassuring intermediate value if you experience extreme blood glucose swings. You can become self-confident and independent from any outsider's perspective and to some extent from an HbA_{1c} measurement from a laboratory. The majority of today's blood glucose meters deliver automatically MBG of the last 14 or 30 days.

The necessary scheduled and planned "diabetes diet" makes me crazy!

The "diabetes diet" for type 1 diabetes was considered necessary because the insulin treatments available were inadequate for replacing insulin deficiency. Regular, portioned and scheduled meals are necessary with conventional insulin therapy (2-3 daily insulin injections) to avoid hypoglycemia. However, if you inject insulin with each meal, you can freely choose the content, quantity and timing of your meals. Regardless of the type of insulin, you must always know the carbohydrate content of your food to select an appropriate insulin dose.

Do I need a carbohydrate exchange list?

It's helpful to learn carbohydrate counting. A carbohydrate exchange list is a table of foods with approximately the same amount of carbohydrates. Foods on this list can be "exchanged" or traded for one another. For example, one slice of bread equals one-half cup of cereal. You need to know which foods are rich in carbohydrates because carbohydrates influence blood glucose values more than protein or fat. Foods that are rich in carbohydrates include breads, pasta, fruits, and most dairy products. In the US, an easy way to discover the carbohydrate content of the food you select is to use the FDA (Food and Drug Administration) food label on the back side of the package. In Europe, packaged foods must have a nutrition guide on the label. The serving size and carbohydrate content is clearly listed at the end of this manual along with other relevant information such as fat and calories for example.

What is a carbohydrate choice?

A carbohydrate choice is a measure of carbohydrates. It is the quantity of food containing 15 grams of carbohydrate in the USA or 12 grams of carbohydrate in Europe. One carbohydrate choice corresponds to approximately one thin slice of bread or half of a medium sized roll. Values can be found in a carbohydrate exchange list or on the FDA label. It is of paramount importance to learn to estimate visually the carbohydrate content of foods. For example, 1 carbohydrate choice or "carb" equals half a large banana, or a small apple, or 1 cup of milk, or 1 thin slice of bread, etc. It's not meaningful to weigh the carbohydrates to the last gram exactly. Your blood glucose varies often regardless of the number of grams in each carbohydrate. Some carbohydrate food choices may increase your blood glucose rapidly, such as rolls, potatoes, fruits or juices, whereas other carbohydrates increase blood glucose more gradually because they are digested more slowly, such as lentils, legumes, starchy vegetables like corn, ice cream, and pasta. The absorption rate of carbohydrates influences the insulin requirement for a particular meal. The slower the absorption of carbohydrates into the intestinal tract is, the lower your prandial insulin requirement will be. We shall return to this topic when you'll be learning how to choose your insulin dosage independently. At the end of this book is a table of "*glycemic index*" reporting the effects of different carbohydrate foods on blood glucose levels.

Am I allowed to eat anything with FIT, even sweet things?

You can eat what you like if you are able to estimate accurately the insulin dose necessary to cover that choice. The FIT program empowers you to make the right decisions about the necessary amount of insulin, the necessary time interval for a meal and the right insulin delivery procedure. It is sometimes difficult to judge how many carbohydrates are in a meal and match that to an appropriate amount of insulin. Sweets are considered "difficult" because mistakes in insulin doses can lead to extremely high or low blood glucose after the meal. "Diabetic" foods contain sugar substitutes (e.g., fructose) that are "utilized" in part independently of insulin: The prandial insulin requirement is lower for these foods.

Is counting carbohydrates accurate enough, or should I also consider calories?

You need to take calories into consideration if you are overweight or prone to being overweight. A *calorie* is a measure of the energy content of food. Foods rich in calories will obviously make you fat when the energy intake exceeds your energy output. Please remember that urinary glucose loss is rare with properly applied Functional Insulin Treatment, FIT. If you experience high urinary glucose values on conventional insulin treatment because of insufficient blood glucose control, part of the carbohydrates you eat, i.e. glucose, are flushed into your urine. This will no longer happen with a more precise diabetes management. Thus, you may put on weight when your glycemic control is better managed because your body will store excess energy as fat. To avoid this, you may need to be careful about the foods you consume or eat less.

How can I do that? I always feel hungry!

If your body weight is normal or below normal, you can eat whatever you want as long as you take the right amount of insulin. To avoid gaining weight, you need to take fat content into consideration. Fat contains the most calories: one gram of fat contains 9 calories compared to 1 gram of protein or carbohydrate which contains only 4 calories each.

I always thought carbohydrates contained the most calories.

This is not correct. You may be confusing fat and carbohydrate calories with insulin requirements. Carbohydrates require much more insulin than proteins or fats. Insulin requirements for proteins or fats are extremely low and can generally be disregarded when eating balanced meals which include carbohydrate-rich foods.

Do I need to avoid fat if I do not want to put on weight?

You need to reduce fat consumption to a minimum. Fatty foods pose a higher threat of cardiovascular risk and contain many calories (9 calories per gram!). You may also need to limit your consumption of alcohol: one gram of alcohol contains 7 calories.

How do you estimate the calorie content of an entire meal?

For carbs it is easy: if 1 gram of carbohydrate contains 4 calories, how many calories does one carbohydrate choice have (1 choice is 15 grams)?

4 x 15 = 60...

Round off the estimated values you already know for carbohydrate-rich foods which also contain protein (like breads, cereals and potatoes) to 70 kcal. E.g., one slice of bread: 15 grams of carbohydrates multiplied by 4 calories each equals 60 calories plus 3 grams of protein is about 70 kcal altogether. Only fruit is a pure carbohydrate, 15 grams x 4 kcal = 60 kcal.

One roll is about 2 carbohydrate choices. Does it have about 140 calories?

That's correct. This rough estimate of calories for cereal products is sufficient.

And what about calorie content of protein foods?

Please note, 100 grams of meat (one portion) corresponds to approximately 150-400 calories. 100g lean meat will contain less than 200 kcal, "usual" sausage, bacon, etc. contain about 250-300 kcal, and fatty products will contain more than 400 kcal/100g.

So, what about fat?

It is easy to calculate the calories of pure fat such as oil or butter. Fat is not water-soluble, 1 tablespoon "light" (emulsion of butter with water) = approximately 14 g, has about 6 grams of fat. This means that 1 tablespoon contains approximately 60 calories. This value is the same for different kinds of fat: from polyunsaturated and monounsaturated vegetable fats like margarine and olive oil to the saturated animal fats like butter.

This morning I ate a buttered roll with about 1 tablespoon of butter (10g) and 2 slices of ham (2 x 25g). I would count 140 calories (roll) plus 90 calories for the butter (10 x 9) plus 2 slices of ham (50g corresponds to approximately 150 calories). Altogether, I consumed approximately 380 calories. Is that right?

Correct. To make things easier, you might round off to 400 kcal. This is how to estimate the number of calories.

Do you also have to inject insulin for protein and fat?

If the meal consists mainly of carbohydrates like bread and pasta, it is sufficient if you calculate the insulin solely for the carbohydrate. Fat and protein slow down the absorption of carbohydrates. With a mixed meal, when carbohydrates are "covered" with protein and fat, the insulin requirement for this meal is the same or maybe even less than it would be for bread alone without the addition of protein and fat. If you only eat large quantities of non-carbohydrate foods (a lot of protein, e.g. pork steak with a lettuce and tomato salad) you can expect your blood glucose to increase slowly if you do not inject any meal-related insulin at all.

Do I need to inject insulin for non-carbohydrate foods if I eat a low carb meal?

You will probably need some insulin for a large meal without carbs because your liver will make glucose from other non-carbohydrate sources. Of course, fat and protein foods require a much, much lower dosage of insulin. We will return to this topic later.

What types of insulin are the best?

It depends on the strategy of insulin treatment. Every type of insulin is good when used according to its action. Origin, composition and above all the effect of the individual types of insulin determine which preparation should be chosen for a particular aim and/or therapeutic regimen.

How do drug companies obtain insulin?

Historically, from the pancreas of cows or pigs. Pig insulin is more similar to human insulin. Today, recombinant DNA technology is a widely accepted process to develop synthetic human insulin. In this technology, modified bacteria are used to produce protein molecules needed e.g. human insulin. In the near future, it is probable that only human insulin will be used in the world. It induces fewer antibodies then animal insulin. In this technology, modified bacteria are used to produce protein molecules needed, e.g. human insulin. If the molecule of insulin is somehow modified to display specific characteristics, that is, to work more rapidly or slowly, it is called insulin "analog".

What is "regular" insulin?

Regular insulin contains no delaying substances. It is a rapid-acting (also known as short-acting) insulin. It was used as early as the 1920's when isolating insulin was first possible. Today, as already mentioned, rapid-acting *insulin analogs* (with slightly changed insulin molecules) work even faster than regular insulin and are becoming therefore more and more popular.

How long does regular insulin last?

In a common dosage, regular insulin works for approximately 4-6 hours but its maximum effect is reached after approximately 1-2 hours. The characteristics of the effect depend greatly on the quantity of the insulin as well as how and where the insulin is injected. Rapid-acting insulin analogs work approximately twice as fast.

How does the quantity of insulin influence the characteristics of the effect?

The more insulin you inject, the longer its effect. The opposite is also true. Less insulin has a shorter effect.

What do you mean by "how and where the insulin is injected"?

Conventional insulin treatment recommends that regular insulin be injected subcutaneously (under the skin) *about 30 minutes before eating.* The conventional, scheduled treatment usually consists of two to three injections daily -- every day, in roughly the same region of the body, e.g., the arm, calf, thigh, buttocks or stomach area. The injection site should be "rotated" every day to reach comparable effects. However, the effect of regular insulin can be accelerated by injecting into a muscle. Muscles are supplied with more blood than fatty tissues, accelerating insulin absorption. Another possibility is to stimulate blood circulation on the skin at the injection site by rapid rubbing or applying heat. We will return to this subject later.

I use regular insulin but I have heard of rapid-acting analogs. How do rapid-acting analogs differ from regular insulin?

Rapid-acting analog insulin was introduced in 1997. It is an insulin "analog", i.e. a somewhat changed molecule of insulin which works very fast, usually twice as fast as regular insulin (examples are: insulin lispro /Eli Lilly, Humalog®/, aspart /NovoNordisk, NovoLog®/ and glulisine /sanofi-aventis, Apidra®/, see Fig.4.2.). Unlike regular insulin, it begins working within a few minutes and can be injected just before or even after eating. It also has a shorter duration of just few hours with an early peak. The shorter duration of action makes insulin analogs very predictable and useful for FIT and specifically for correctional purposes, if your blood glucose is high. Changing the mode of injection (e.g. injecting into muscle rather than fat, or increasing the blood flow on the injection site as necessary for regular insulin) becomes virtually irrelevant with insulin analogs.

What are the differences between delayed-acting (or long-acting) insulin and regular insulin?

First, delayed-acting insulin work longer, as their name implies. Second, they differ according to composition. The long-acting insulin contains other substances like protein bodies or zinc ions. These delaying substances along with insulin create crystals that reduce the solubility of insulin and slow down the delivery of insulin into the blood from the injection site. Third, these additional substances in long-acting insulin create a different "cloudy" appearance. Long-acting insulin separates and so it needs to be rolled or gently shaken before being drawn from the bottle into the needle and injected. Zinc insulin (lente and ultra-lente types) are no more available. Instead, delayed acting insulin analogs, Lantus® and Levemir® have been recently introduced.

Delayed acting insulin analogs have other mechanisms for delaying of their action: Lantus®/ glargine (sanofi-aventis) is sour and will be freed into the blood stream with normalization of pH (sour) at the injection site. Levemir® /insulin detemir (NovoNordisk) binds on plasma protein and that delays its action. Levemir® acts somewhat shorter than Lantus®.

How long does the delayed-acting insulin last?

There are several delayed-acting insulin on the market. The three most important groups (see Schemes, Fig. 4.2) are:

1. **NPH insulin**: This abbreviation means "Neutral Protamine Hagedorn". Hagedorn was the first person to use protamine, a fish protein, for delaying the absorption of insulin in the 1940's. In a usual dosage, NPH insulin works for about 10-18 hours. The peak effect is reached in 4-6 hours. All insulin manufacturers produce NPH insulin.
2. **Insulin analog detemir** (Levemir®, NovoNordisk) acts similar to the previous zinc insulin Lente (not available any more), works longer than NPH insulin and reaches its peak effect somewhat later and longer.
3. **Insulin analog glargine** (Lantus®, sanofi-aventis) acts even longer, but in contrast to the previous zinc insulin Ultralente (crystalline zinc insulin, not available any more) it starts to work quite fast, almost immediately after injection.

These new long acting insulin analogs (glargine /sanofi-aventis: Lantus®/ and insulin detemir /NovoNordisk: Levemir®) have been introduced to the market a couple of years ago. Lantus® profile is quite similar to that of the Ultralente, however it seems to work even a bit longer and starts to work very much earlier as it. It provides a basal insulinization often stable enough even if injected only once per day. The profile of Levemir® resembles more that of Lente or even NPH in small amounts. The delaying principle with Lantus® is sourer than the body fluids so that it gets released only gradually and very slowly after injection while neutralized. Levemir® is the first insulin preparation that bounds to blood proteins; this insulin action delaying principle is completely new and promising.

Which insulin is the best for my fasting requirements with FIT?

With two injections a day, each with approximately the same dosage, you can use any combination of these long-acting insulin as a basal insulin. With tendency to higher morning values of blood glucose -- a frequent finding called "dawn phenomenon" -- Lantus® insulin during the day in combination with and NPH or Levemir® insulin late at night seem to be the best in practice. We will come back to this subject later. Insulin analogs Lantus® or Levemir® cannot be mixed with other insulin preparations in the same syringe. So you can see that individual needs or preferences need to be taken into account while you decide on the type of insulin for your basal requirement.

Which insulin treatment is the best?

The best insulin treatment is what works for the individual. The choice of a particular strategy depends first on *your* goals. The younger you are, the more you need to aim for a near-normal blood glucose. Only then can the late complications of diabetes be delayed or avoided altogether. The older you are when your diabetes is diagnosed, the less likely is that late complications will develop. Understandably, other treatment aims have to be taken into consideration if you have another serious illness in addition to diabetes like cancer, which may shorten your life expectancy. Serious hypoglycemia needs to be avoided at all times but in older people might be even more dangerous.

What are the strategies for insulin treatment?

To make things easy, we differentiate between two types: (1) **conventional** and (2) **functional** insulin treatment. The last and most advanced category allows a near *n*ormal glycemic insulin replacement (FIT) insofar as insulin is used according to certain rules.

And what about the new insulin preparations and the cancer risk?

This is indeed a controversial issue, although it has recently been shown that glargine (Lantus®) is associated with less cancer cases as other insulin preparations, there has been some association for cancer probability and the daily glargine dosage in group of type 2 diabetes patients treated exclusively with this insulin. So they got only basal insulin. It looks like this phenomenon proves an already known fact , that insulin resistance per se (and thus high glargine dose) is associated with somewhat increased cancer risk.

		Insulin Type Delaying Substance	Pharmacokinetics (Examples)	Manufacturer		
				Sanofi-Aventis	Eli Lilly	NovoNordisk
Rapid-acting Insulin		Rapid-acting Analogs	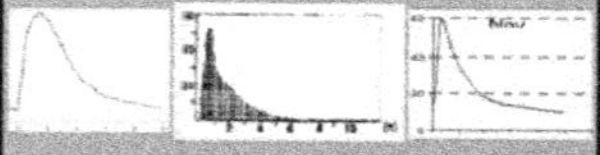 Apidra / Humalog / NovoRapid	Apidra (Insulin glulisine)	Humalog (Insulin lispro)	NovoLog (Insulin aspart)
Rapid-acting Insulin		Regular Insulin	Actrapid HM 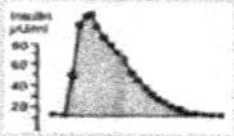	Insuman Rapid	Humulin R (Regular, R)	Novolin R
Delayed-acting Insulin	Slightly delayed-acting	NPH-Type (Protamine)	Insulatard HM 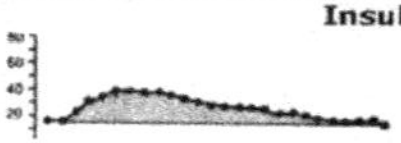	Insuman Basal	Humulin N	Novolin N
Delayed-acting Insulin	Medium long-acting	**Earlier:** **Lente-Type** **(Zinc crystalline + amorph)** 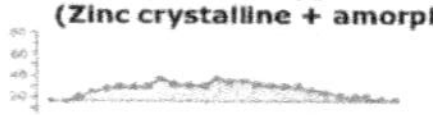	Levemir 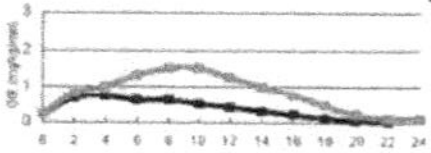			Levemir (Insulin detemir)
Delayed-acting Insulin	Long-acting	**Earlier:** **Ultralente-Type** **(Zinc crystalline)** 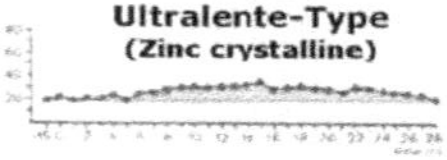	Lantus 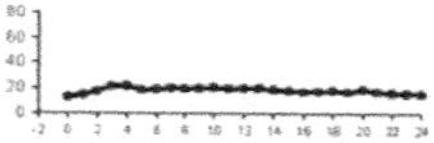 Time (hrs)	Lantus (Insulin glargine)		

Fig. 4.2.: Selection of human insulin and insulin analogs

Notes:

- A large number of mixtures of regular and delayed-acting insulin in varying proportions are commercially available. Since these premixed insulins are not suitable for the purposes of FIT, they are not listed here.
- Rapid-acting insulin analogs, insulin glulisine, lispro and aspart have a considerably faster effect than regular insulin. The activity curves shown here are based on an insulin dosage of 10 units (i.e., less than half the dosage shown in the examples of other insulin types; based on Galloway 1993).
- Long-acting insulin analogs (glargine /Lantus®, sanofi-aventis/ and insulin detemir /Levemir®, NovoNordisk/) were explored regarding the pharmacokinetics in not comparable dosages to other preparations. Levemir-effect curves correspond to a dosage of 0.2 (lower curve) and 0.4 (upper curve) units per kg body weight. Both long-acting insulin analogs which are available on the market cannot be mixed with short-acting insulin.
- Zinc insulin profiles according to Bottermann et al 1985 (in dosage 0.3 U per kg body weight; activity profiles as examples for insulin of NovoNordisk). With the small amounts of insulin per dose resulting from the separate administration of basal, prandial and correctional insulin as required by FIT, the activity duration can be expected to be correspondingly shorter.
- Since 2007 the zinc insulin (type Lente or Ultralente) from the above mentioned manufacturers is neither available on the American nor European markets anymore.

Adapted from Howorka K: Functional Insulin Treatment, 3rd ed., Springer Publishers Berlin, 2009

TREATMENTS:	*Conventional*	*Intensive*	**Functional**
Insulin	Long-acting or pre-mixed insulin.	Short- and long-acting insulin injections: some insulin dose adjustment, no premixed insulin	Numerous injections of short-acting and long-acting insulin or insulin pump
Number of injections per day	1-2	3-4	Varies: average of 4-7
Function: Basis/Bolus	Global coverage of basal (fasting) and prandial (mealtime) insulin requirements.		Clearly separate insulin coverage: • basal • prandial • correcting (hyperglycemia)
Diet	Fixed. Meals and schedule are adapted to the effects of insulin: time, quantity and content.	Slightly more flexible because regular insulin is used to cover meals.	Liberated. Meal content, quantity & timing to suit the individual; insulin is delivered when needed.
Blood glucose self-checks	Rare or seldom blood glucose glucose checks.	4 times daily blood glucose checks.	4-6 times or more daily blood glucose checks are necessary to normalize HbA_{1c}.
Consequences of blood glucose self-checks	Little independence, strong reliance on physician intervention for insulin adjustments No immediate correction of hyperglycemia. Hypoglycemia corrected with carbohydrates.	Delayed independence. Changes in insulin dosage when the same glucose pattern arises over 2-3 days.	Total independence. Changing insulin dosage is immediate (algorithms) using blood glucose targets.
Training	None or limited.	Diabetes training for independent insulin dosage adaptation. Most centers provide approx. 20-25 hours of one-on-one or group training	Intensive group education over a minimum of 40 hours with specific modules: hypoglycemia, hyperglycemia, nephropathy, neuropathy, etc. FIT algorithms in theory and practice: insulin games • intentionally increase and decrease BG • experimental fasting • experimental „sinning" • learning how to critically test and adapt algorithms of functional insulin use.

Fig. 4.3: Principles of insulin therapy for insulin-dependent diabetes

These rules govern whether insulin is used to compensate for eating or for fasting or for correcting blood glucose values outside the target level. As a reminder, Functional Insulin Treatment is realized with several daily insulin injections or with an insulin infusion pump. The prerequisite is that you have acquired relevant training and knowledge and are capable of making safe, accurate choices for dosage. Blood glucose self-monitoring at least four times every day and a daily balance of insulin and food consumption is indispensable to achieve near-normal blood glucose and HbA_{1c} results.

"Intensified" Conventional Insulin Therapy	Functional Insulin Treatment
Advantages	
• Only 2 or 3 injections • Part of the responsibility for treatment can be given to the doctor (an advantage?) • Only 5-7 "painful seconds" for blood tests / injections per day	• Good to excellent metabolic control • Independent diet regarding: (1) quantity (2) timing (3) frequency (4) content of meal • Flexible insulin injection times • Deliberate influence on blood glucose • Transparency of metabolic processes (functionally separated insulin use) • Far-reaching independence from physician and prescription • Self-responsibility
Disadvantages	
• Required to adhere to a strict dietary plan re: (1) quantity (2) timing (3) content (4) frequency of meals • Fixed injection times • Fewer possibilities to monitor blood glucose • Danger of hypoglycemia when postponing meals • Loss of spontaneity • Poor to good glycemic control	• In total, at least 10-20 "painful seconds" for blood tests and injections per day • If pump is used, it must be worn on the body at all times

Fig. 4.4: Advantages and disadvantages of different types of treatments

Contrary to this, there is the usual "conventional" insulin therapy. This therapy uses insulin dosages always prescribed at the same time of the day in accordance with a planned dietary schedule. Day-to-day variations in the quantity or timing of food are almost impossible. If the therapeutic goal is to normalize your glycemic control, to extend your life expectancy, to avoid late complications, than three or more insulin injections a day are an absolute minimum, even with conventional insulin treatment. In this case, we avoid using pre-mixed insulin, but recommend that you independently select the composition of your morning and evening injections, each consisting of rapid- and delayed-acting insulin. At least four blood glucose self-checks should be carried out every day. The term "intensified" insulin therapy is often used when the patient carries out self-checking measurements and independently "adapts" insulin dosage accordingly. In type 1 diabetes, such "intensified" conventional insulin therapy should be the absolute minimum treatment recommended.

Does that mean that insulin injections twice daily without blood glucose self-monitoring are really not enough?

Yes, especially if your therapeutic goal is to achieve long-term near normal blood glucose. However, many claim insulin treatment without accompanying self-monitoring may be the only feasible therapy for some older people with diabetes.
Characteristics of individual strategies for insulin treatment with conventional insulin therapy, intensified insulin therapy, and FIT are compared in the Fig. 4.3. Conventional insulin therapy in the "classical" sense with scheduled meals, scheduled injections and no self-adaptation of dosage cannot be recommended anymore, since self-monitoring has no implications to the patient. The treatment should be "intensified" and more flexible, without premixed insulin to make the patient capable of making informed, appropriate decisions on insulin dosage and adaptation.

Can anyone avoid glucose self-monitoring?

The more varied your lifestyle, and the more you vary food intake and physical activity from day-to-day, the more helpful blood glucose tests are. Only experience can tell you how often you should measure your blood glucose so that your HbA_{1c} remains close to normal. Insulin requirements vary greatly from one day to the next and are not always predictable, even under constant nutritional conditions. Virtually, the only way to overcome this is to immediately and accurately correct blood glucose levels outside your target. That remains valid also with contemporary trials of continuous blood glucose monitoring. Think about the pros and cons of each type of treatment before selecting a particular strategy of insulin treatment (see Fig. 4.4.). Conventional insulin therapy without self-monitoring and insulin self-adjustment is not included in this diagram for obvious reasons. Here, we are comparing the functional use of insulin or FIT with the less differentiated forms of treatment where insulin is administered globally at the same time for prandial and basal functions as described with Sklyer algorithms see Fig 2.1. In this case, the metabolic processes are not transparent; nor are patients empowered to influence their blood glucose values.

I can already see that FIT has many advantages but I am not sure if I can cope with the disadvantages of this method.

The biggest advantage or disadvantage, depending on your perspective, is the responsibility for yourself. You will need to manage your metabolic control. Would you choose to leave your blood glucose at 300 mg/dl when you could immediately reduce it to 100 mg/dl? The rewards are freedom and very good control of your blood glucose with the consequence of frequent blood glucose tests and injections or an insulin infusion pump.

Yes, OK! Being responsible for myself frees me from such obligations as waiting for the doctor to "fix" my regimen, following rigid prescriptions, eating at particular times, particular diets and traveling frustrations. I like those advantages. Still, I cannot believe how many FIT times I will need to jab myself: 4 for self-monitoring and maybe 6 for injections. At least 10 "pricks" every day!

That is correct. Only when you accept these 10 seconds of pain a day and a time expenditure of about 5 minutes every day, will the advantages be realistic for you. You have to make the decision on your own. Don't let anyone influence you. Also, don't believe that an insulin pump or an insulin pen (a pen-shaped device for injecting insulin) will automatically manage everything for you. An insulin pump or an insulin pen is only as valuable as the knowledge or the actions of its user. They do not guarantee good control by themselves. What is important are your own decisions based on self-monitoring, carb counting and the quantity of insulin you inject each time -- these are the factors that count.

If I decide to opt for FIT, does a portable, controllable insulin pump have any advantages compared with multiple daily injections?

Yes, in so far as your basal insulin is substituted with a continuous infusion of small amounts of rapid-acting insulin, which is often even more stable than two daily injections of delayed-acting insulin. On the other hand, by using injections at mealtimes, you can intentionally increase the absorption rate of your regular insulin by injecting into muscles and/or rubbing the site. With pump therapy, this would require various kinds of bolus.

With insulin pumps, subcutaneous insulin administration with a catheter under the skin becomes more and more popular. Forms of intravenous (into the vein) or intraperitoneal (abdominal) infusions are almost no longer used. Later in this book we will go into technical details of dealing with the pump, changing catheters, alarm functions, and so forth, which differ according to the model (see the chapter on pumps). The introduction of continuous glucose monitoring, although at present to expensive for coverage by insurance, could facilitate the FIT with pumps even more. Now we will concentrate on the principles of insulin dosage. These are basically identical in FIT regardless of how insulin is administered, by injections or a pump (apart from the fact that insulin requirements in continuous insulin infusion are somewhat less). Don't immediately accept it or reject it. You might consider trying the pump for a couple of weeks and then making a decision that is right for you. In any case, it is not advisable for you to rely totally on devices like pens or pump to deliver your insulin as sooner or later they may fail. Therefore, it's a good idea to keep in good practice of injecting insulin with older, traditional syringes. In any case it is good to consider potentially higher risk of technical damage and failure with complex technologies.

Test Questions on Phase One - Basic Diabetes Training

Name: .... Date:...................................

All questions have one or more correct answers:

A) One carbohydrate choice corresponds to

		true	*false*
1.	1 apple	☐	☐
2.	1 big banana	☐	☐
3.	1 small grapefruit	☐	☐
4.	1 egg sized potato	☐	☐
5.	2 tablespoons of mashed potatoes	☐	☐
6.	1 cup (250g, 8 oz.) of whole milk	☐	☐
7.	2 cups (500g, 16 oz.) of skimmed milk	☐	☐
8.	1 cup (250g, 8 oz.) of yogurt	☐	☐
9.	1 slice of brown bread	☐	☐
10.	2 slices of whole wheat bread	☐	☐
11.	1 small ladle of cooked pasta	☐	☐
12.	1 small glass (125g, 4 oz.) orange juice	☐	☐
13.	5 or 6 saltine crackers	☐	☐

B) The following vegetables do not have to be calculated into the choice of insulin dosage, even in large quantities:

		true	*false*
1.	cucumbers	☐	☐
2.	peas	☐	☐
3.	sauerkraut	☐	☐
4.	corn	☐	☐
5.	lentils or beans	☐	☐

C) Three tablespoons (about 20 grams) butter corresponds to approximately:

		true	*false*
1.	100 calories	☐	☐
2.	170 calories	☐	☐
3.	310 calories	☐	☐

D) One carbohydrate choice of bread and cereal products usually contains:

		true	*false*
1.	20 calories	☐	☐
2.	70 calories	☐	☐
3.	150 calories	☐	☐

E) 100g of meat contains:

		true	*false*
1.	150-200 calories, if it is lean	☐	☐
2.	300-400 calories, if it is fatty	☐	☐
3.	on average, 250-300 calories	☐	☐

F) For a younger person who has had type 1 diabetes for three years and holds a job, the following types of treatment are suitable:

		true	*false*
1.	Levemir® insulin once a day and no self-monitoring whatsoever	☐	☐
2.	Premixed insulin twice a day and a urine glucose check every other day	☐	☐
3.	Twice daily mixed insulin and scheduled food intake, regular check-ups with a general practitioner and no self monitoring	☐	☐
4.	Mornings and evenings, dosage choice of delayed-acting and rapid-acting insulin; self-monitoring blood glucose several times a day, scheduled food intake	☐	☐
5.	Delayed-acting insulin twice daily, rapid-acting insulin several times a day, flexible diet, blood glucose self-monitoring several times a day	☐	☐
6.	Controlled insulin delivery via a pump, flexible diet, daily self-monitoring of fasting blood glucose	☐	☐

G) In the case of adult type 1 diabetes patients able to self-adjust their insulin dosage, achieving a near-normal HbA_{1c} is most likely with the following types of self-monitoring:

		true	*false*
1.	A blood glucose profile consisting of 7 values performed once every week	☐	☐
2.	Blood glucose self-monitoring twice daily	☐	☐
3.	Fasting blood glucose values every day	☐	☐
4.	Blood glucose self-monitoring 6 times a day, every third day	☐	☐
5.	Blood glucose self-monitoring at least 4 times daily	☐	☐
6.	Urine glucose monitoring 3 times daily	☐	☐

H) A lack of insulin definitely exists when:

		true	*false*
1.	Ketones are present in the urine and the urinary glucose is negative	☐	☐
2.	Ketones are present in the urine and the blood glucose is 190 mg/dl	☐	☐
3.	Ketones are ++ positive in the urine and the blood glucose is 90 mg/dl	☐	☐

I) The delayed acting insulin listed here are in the right order, increasing or decreasing according to their duration:

		true	*false*
1.	Lantus®, NPH insulin, Levemir®	☐	☐
2.	NPH insulin, Levemir®, Lantus®	☐	☐
3.	Levemir®, premixed insulin, Lantus®	☐	☐

J) To establish the basal rate for Functional Insulin Treatment, the following insulin is suitable:

		true	*false*
1.	Lantus® twice daily	☐	☐
2.	Lantus® once daily	☐	☐
3.	NPH insulin twice daily	☐	☐
4.	Premixed insulin twice daily	☐	☐
5.	Levemir® twice daily	☐	☐

K) A urine glucose test indirectly informs us about the blood glucose value:

		true	*false*
1.	in the last five hours	☐	☐
2.	between the last two urinations	☐	☐
3.	in the last hour before urinating	☐	☐

L) Indicate whether the following sentences are true or false:

		true	*false*
1.	Glucagon increases blood glucose.	☐	☐
2.	The only function of the pancreas is to produce insulin and glucagon.	☐	☐
3.	When you drink alcohol, glucose production in the liver decreases.	☐	☐
4.	When someone with diabetes drinks alcohol, they must inject more insulin.	☐	☐
5.	Glycogen and glucagon are the same.	☐	☐
6.	Physical exercise reduces the effect of insulin.	☐	☐
7.	People with diabetes are always fat.	☐	☐
8.	Type 1 diabetes always occurs before the age of 25 years old.	☐	☐
9.	The thirsty feeling you have with high blood glucose is due to a flushing effect of urinary glucose.	☐	☐
10.	Acetone in the urine demonstrates that the body is obtaining most of its energy from fat stores.	☐	☐
11.	With type 2 diabetes, treatment with tablets is *always* the chosen therapy.	☐	☐
12.	The late complications of diabetes develop only in people with type 1 diabetes.	☐	☐
13.	Even severe late complications of diabetes can be cured if the metabolic (glycemic) control is very good.	☐	☐

	true	*false*
14. Microangiopathies are the characteristic changes in the small blood vessels in diabetes.	☐	☐
15. Good metabolic control has no value if late complications already exist.	☐	☐
16. HbA_{1c} describes the quality of glycemic control during the last 6-8 weeks prior to the blood test.	☐	☐
17. Diabetes patients have a lack of HbA_{1c}..	☐	☐
18. 1 gram of fat = 9 calories.	☐	☐
19. 1 gram of protein = 4 calories.	☐	☐
20. 1 gram carbohydrate = 9 calories.	☐	☐
21. 1 gram alcohol = 4 calories.	☐	☐
22. 1 gram water = 1 calorie.	☐	☐
23. FIT is a new type of oral insulin therapy.	☐	☐
24. Insulin pens are absolutely necessary for FIT.	☐	☐
25. Good metabolic control can only be achieved today when people with diabetes lead a rigid lifestyle.	☐	☐
26. Late complications of diabetes can be postponed or avoided with good metabolic (glycemic) control.	☐	☐
27. Insulin must always be stored in the refrigerator, even a bottle opened recently.	☐	☐
28. Disinfecting skin before injecting insulin is always necessary.	☐	☐
29. FIT requires individually tailored dosages of insulin for eating, fasting, and correcting blood glucose.	☐	☐
29. Carbohydrate counting and dietary knowledge are not necessary with FIT.	☐	☐
30. Today, in the practical life of a person with type 1 diabetes, insulin production can only be mimicked with functional use of insulin.	☐	☐
31. FIT can be achieved with a pump as well as multiple daily injections.	☐	☐
32. The goals of FIT are good glycemic control and a high quality of life (flexible lifestyle).	☐	☐
33. Checking HbA_{1c} levels once a year is enough with insulin-dependent diabetes.	☐	☐
34. The target mean blood glucose (MBG) with type 1 diabetes is 90 mg/dl.	☐	☐
36. You need never measure blood glucose after eating.	☐	☐
37. Long-acting insulin is always used as prandial insulin on FIT.	☐	☐
38. Carbohydrates contain more calories than fat.	☐	☐
39. People with diabetes need to eat the least amount of carbohydrates possible.	☐	☐

	true	*false*
40. Dietetic foods with sorbitol, xylitol and fructose require just as much insulin as foods sweetened with sugar.	□	□
41. Insulin should never be injected into the muscles.	□	□
42. Only human insulin should be used on FIT	□	□
43. FIT can be achieved with twice daily injections of rapid-acting and long-acting insulin.	□	□
44. Self-monitoring can be completely avoided if a person with type 1 diabetes goes to the doctor regularly.	□	□
45. Intensified insulin therapy requires a self-adjustment of insulin dosage according to results of daily self-monitoring.	□	□

Further reading:

If you have 6 or more wrong answers to questions on "basic diabetes training", we recommend that you take part in an appropriate diabetes training seminar and read one of the classical books on intensified conventional insulin treatment before you start with FIT training.

We can recommend, e.g.:

- Walsh J, Roberts R, Varma Ch., Bailey T: Using Insulin. Torrey Pines Press 2003
- Warshaw H.S., Pape J: Real life guide to diabetes. ADA Bookstore, 2009

3. Phase Two - FIT Training

If I decide to follow the FIT method now, do I have to follow it for the rest of my life?

No you don't but, you can always return to your earlier treatment with intermediary insulin. In my experience however, this is hardly ever the case... Every type of treatment has advantages and disadvantages. You could opt either for treatment with two to three injections and a scheduled diet (even here, self-monitoring is essential) or for FIT with advantages of a flexible diet and better control, using a pump or injecting more frequently.

Do I need to be admitted to the hospital to learn FIT?

No, not at all. We do the training on an outpatient basis. Initially a residential course for a couple of days has proven successful, to protect you during the learning phase and to shield you from the chaos of everyday life. Later on, our experiences have clearly shown that the education can be carried out successfully in an outpatient setting. Basic diabetes training and FIT module may occur at different times. This allows you to choose an appropriate time to participate in the courses.

I have decided to start FIT using several injections. Can I start immediately?

Yes. FIT is based on functionally separated use of insulin for:
1. Fasting
2. Eating
3. Correcting blood glucose values outside the target range.

All of the rules that describe how you replace fasting insulin, prandial insulin, and correcting insulin are called **FIT algorithms of insulin dosage**. They give concrete answers to the questions summarized in Chapter 3, Fig. 3.1. (Note: Please read these questions once more now). Here are some tips about the organization of the learning phase for FIT:

1. Try to measure your blood glucose as frequently as possible during the course to develop a feel for what is happening with your blood glucose when you eat or inject insulin. It is necessary to take approximately 10 blood glucose measurements every day, *that much only* during the learning phase. You need to measure blood glucose:
   * before going to sleep
   * at 2 or 3 a.m.
   * fasting -- in the morning
   * prior to eating, and
   * about 1-2 hours after eating

   Later, on your own, after you have developed an awareness of your own blood glucose and have a "feel for insulin replacement", 4-6 daily measurements are usually enough to near-normalize HbA_{1c}, only very few patients need measurements more often.
2. Always be sure to check the accuracy of your blood glucose self-measurements. Occasionally, you need to measure your blood glucose and compare it with a measurement at the doctor's office.
3. Your diabetes physician or advisor should examine your technique of blood glucose measuring, comment on it, and if possible simplify it. Ask your diabetes advisor if your technique for testing is simple and accurate enough.
4. Also, ask your physician or diabetes advisor for their opinion regarding your injection technique. Use only syringes with welded-in needles that can be used several times. Insulin syringes for lower insulin dosages are useful with daily insulin requirements between 30 and 60 units. In particular, it is possible to use half units with small dosages of insulin. Rarely are more than 10-15 insulin units injected at any given time, usually less with the insulin requirement already mentioned under FIT.
5. Lancets for blood tests and needles for injecting can be re-used. It is important for your own safety that you do not share your devices with other people. Always dispose of your sharps in an approved manner. It will protect yourself and others from communicable diseases.
6. You may also want to check your urinary glucose and acetone secretion at least several times a day during the group education course for FIT. This has proven very important during the learning phase to develop a feel for your metabolic control. If possible, find out about the excretion of glucose in your 24-hour urine sample. In general, none or only a few grams/day characterize good glycemic control. Measurement of urinary glucose does not allow appropriate blood glucose corrections, so it is not used in FIT for this purpose

7. Document the results of your self-monitoring on your record sheets. Fig. 5.1 shows an example for documenting your blood glucose, insulin dosage, and food intake. The most important items on this FIT record sheet are the sums, the "daily balance", which need to be calculated every day: The daily total of insulin units injected, carbohydrates consumed, and your mean blood glucose (MBG) from each day combined to achieve the weekly MBG.
8. To ensure insulin dosing accuracy *during the learning phase (structured FIT training)*, use syringes rather than insulin pens or a pump. You need to learn many new things and the insulin pen and/or a pump, which are definitely an important option for the future, may distract you from what is really important during the learning phase. Use the pen or pump only when you clearly understand your insulin dosage guidelines, for now you should focus rather on your blood glucose course, not the device.

How are my FIT algorithms calculated for insulin dosage?

There are two facts that are particularly important here:
1. Your current daily insulin requirements, and
2. Your glycemic control up to now.

Of course, your diet, frequency of hypoglycemia, and possibly your urinary glucose and acetone values need to be taken into account. Assuming that you have an insulin requirement of about 40-50 insulin units a day on your usual diet, you can use the values of the insulin production rate of a person without diabetes as a model for your algorithms:

1. The **basal insulin** may amount to approximately 20 units per day and you could replace this by injecting long-acting insulin twice a day with 10 to 12 units. Some units of rapid-acting insulin may be added to the basal insulin in the morning, because the insulin requirement is at its highest then. We will return to this later.
2. The **prandial (meal-related) insulin** may frequently be approximately 1-2 units per carbohydrate choice. To replace the prandial insulin, only regular insulin or rapid-acting analogs insulin can be used.
3. **Correction insulin:** 1 unit of rapid-acting insulin may often be expected to decrease the blood glucose by 40mg/dl (with the aforementioned daily insulin requirement of 40-50 units) for a person with diabetes under "basal conditions" and stable blood glucose levels. One carbohydrate choice (15g of quickly absorbed carbohydrates) increases the blood glucose by about 50 mg/dl.

There are simple equations to derive your basal, prandial and correction FIT algorithms from your total daily insulin need. You will find them at the end of the book. Your physician will help you with estimation of your initial FIT algorithms at the end of the Basic Diabetes Education.

A FIT record sheet is for recording insulin dosage, blood glucose values, and food intake. The common record books generally available for „insulin dosage adjustment“ for predetermined fixed doses are not suitable for FIT because they presuppose a regular diet and fixed schedule. The row for long acting/ basal insulin can be used in pump therapy to note the basal dose of short acting insulin per hour (once per page, e.g. Mondays only as on example, is enough).

The last line of the day is free for personal notes: for example calorie intake for those who control the energy intake, separate remarks for differentiating the amount of correctional from that of prandial insulin (usually written in the row for rapid-acting insulin), so called carb factors in case the prandial insulin dose varies with the time of the day (rarely needed with correct basal substitution compensating circadian, time-related variations of insulin requirement), or controlling/recording undesirable behavior, i.e. number of cigarettes, amount of alcohol, in pregnancy it can be used for monitoring and recording child movements during the last trimenon. The algorithm for insulin dosage for protein and fat is only valid for carb-free meals. The most important last column on the record sheet shows the balance of the day (sums for insulin used, mean blood glucose [MBG] and total carbohydrates consumed for 24 hours). The graphic representation of blood sugar (as diagram) usually gives no extra information. FIT record sheet can be ordered and downloaded anytime directly from our website www.diabetesFIT.org or via www.lulu.com. You will find there some additional material concerning diabetes (i.e. books, lectures, workshop schedules and the latest news and events), as well.

INTERNATIONAL RESEARCH GROUP ON FUNCTIONAL INSULIN TREATMENT
Medical University Vienna
kinga.howorka@meduniwien.ac.at
www.diabetesFIT.org

Name :..

Birth date:.....................Phone:...........
Address:...
E-mail..
Diabetes.since:.................Wt.:.............
FIT since:........... with O injections O pump

I BASAL (fasting insulin): AM ..../........................U N PM ....U S PRANDIAL: for 1 carb choice (15g) = ...U U	**Target for correction of aberrant BG values:** Fasting/pre-meal: 100 mg/dl (or ....) After meals: 1h<180 (or............), 2h< 140 mg/dl **Target range for MBG: from .... to .... mg/dl**

L CORRECTION: 1 U rapid ins. lowers my BG by approx.- ....; 1 carb raises my BG by approx.+ ....mg/dl
I EXAMPLE: carb choices/CHO: ....
N insulin (U): ....

TIME		1	2	3	4	5	6	7	8	9	10	11	12	1	2	3	4	5	6	7	8	9	10	11	12	Total Daily Dose
							AM													PM						
MON	Basal																									
	Bolus																									
	BG																									MBG
	Carbs/CHO																									
	Comments																									
TUE	Basal																									
	Bolus																									
	BG																									MBG
	Carbs/CHO																									
	Comments																									
WED	Basal																									
	Bolus																									
	BG																									MBG
	Carbs/CHO																									
	Comments																									
TIME		1	2	3	4	5	6	7	8	9	10	11	12	1	2	3	4	5	6	7	8	9	10	11	12	Total Daily Dose
THU	Basal																									
	Bolus																									
	BG																									MBG
	Carbs/CHO																									
	Comments																									
FRI	Basal																									
	Bolus																									
	BG																									MBG
	Carbs/CHO																									
	Comments																									
SAT	Basal																									
	Bolus																									
	BG																									MBG
	Carbs/CHO																									
	Comments																									
SUN	Basal																									
	Bolus																									
	BG																									MBG
	Carbs/CHO																									
	Comments																									
TIME		1	2	3	4	5	6	7	8	9	10	11	12	1	2	3	4	5	6	7	8	9	10	11	12	
																										weekly MBG

Fig. 5.1: FIT Record sheet

And what if someone has a much higher daily insulin requirement?

The algorithms for basal insulin and prandial insulin would need to be increased proportionally. If someone has a much lower daily insulin requirement than 40 units per day, these algorithms need to be decreased in a proportional way. We work on these algorithms as a group during class with professional supervision.

How does my glycemic state up to now affect the values of my algorithms?

It only indicates whether your insulin dosage was correct up to now. The insulin dosage was too low if people discharge a lot of glucose in their urine and have consistently high MBGs, or often test positive for acetone. On the other hand, for people who have frequent hypoglycemia and very low daily mean blood glucose values (under 120 mg/dl) we can suspect that their total daily dose is too high. Try to judge your metabolic control up to now before starting FIT. To do this, calculate your daily MBG over the last couple of days always including values after eating. To evaluate your control under conventional treatment, use the following guidelines:

MBG higher than 200 mg/dl = too high
160-200 mg/dl = unsatisfactory
130-160 mg/dl = acceptable
below 130 mg/dl = too low, suspicion of too much insulin, unacceptable for those with history of hypoglycemia with unconsciousness

What does the urinary glucose measurement show?

If you have a "usual" kidney threshold for glucose (approx. 180-250 mg/dl) you should remain almost free of urinary glucose with an "acceptable" blood glucose average. The significance of any one urinary glucose measurement may be increased when the urinary glucose discharge over 24 hours is determined:

- A urinary glucose discharge up to a few grams a day often indicates good metabolic control;
- up to 20 grams per 24 hours, insufficient metabolic control;
- and excretion of more than 20 grams per day often corresponds to poor glycemic control.

You can assume that your insulin dosage is too low when positive acetone is found in the urine. Your HbA_{1c} value is better judgment concerning your glycemic control up to now. Your HbA_{1c} is the glycated hemoglobin measurement which was already discussed.

Additionally, to judge present glycemic control, you need to consider how often you experience hypoglycemia and fluctuations in your blood glucose. Under conventional treatment, average blood glucose of 120 or 130 mg/dl is not considered to be a safe treatment goal for people with type 1 diabetes if hypoglycemia is frequent or severe. Review blood glucose values from the last couple of days. How often did values under 60 mg/dl occur? Which blood glucose level do you perceive as being in a state of hypoglycemia? How often do you experience hypoglycemia?

Should I never experience hypoglycemia at all?

Mild hypoglycemia, which can be treated immediately, is not so important and for most people is somehow unavoidable. Thus, it is important to always carry glucose tablets or concentrated sweets with you in order to be safe. However, serious (or severe) hypoglycemia, with loss of consciousness needs to be avoided. We will discuss this in more detail in the chapter about hypoglycemia,. At the moment, your past experiences with hypoglycemia are important because they determine your choice of glycemic target for correcting hyperglycemia and an MBG target area suitable for you.

The record sheet states that values of 100 mg/dl fasting pre-meal and 160 mg/dl after eating have been chosen as target points for correction.

These blood glucose correction targets are only applicable to people with diabetes who have not experienced severe hypoglycemia with loss of consciousness and who are quick to notice values under 50 mg/dl. However, if you have had repeated, serious hypoglycemic episodes in the past, you will need to choose a slightly higher target point for correction: 120 mg/dl for fasting/pre-meal and up to 180 or even up to 200 mg/dl after eating.

The target range for the daily average (MBG) needs to be distinguished from blood glucose target points for correction. If you aim for 100mg/dl fasting/pre-meal and before eating and about 180 mg/dl after eating, that will usually correspond with a target area for mean blood glucose (MBG) between 110 and 160 mg/dl. High risk of serious hypoglycemia (loss of consciousness in the past), means that you need to aim for higher MBG target levels (140-200 mg/dl).

Moreover, you need to aim for a higher blood glucose target area (at least during the learning phase), if:

1. You have severe late complications of diabetes (e.g., serious retinal changes) and if your metabolic control until now has not been optimal. In this case, you need to avoid an abrupt decrease in the mean blood glucose and allow your body to adapt with time. Speak to your diabetes physician about whether it will be possible for you to subsequently lower the blood glucose target zone over time.
2. You tend toward perfectionism and are very ambitious. We hold nothing against ambition but we have noticed that those who are ambitious often take somewhat more insulin than necessary. If you tend toward perfectionism, it is advisable to select a higher blood glucose target range to consciously avoid possible insulin overdoses. Later, after you have had more experience with FIT, you will be able to consider whether your target zone could be lowered safely.

You and your diabetes physician can establish the algorithms of insulin usage after considering all of the aspects of your past and present insulin dosage. Note these on your record sheet. Also, calculate a temporary insulin dosage for the first day, in case your blood glucose always stays at the target point for correction (which is quite unlikely).

So, if I apply these figures, can I eat what I like and bring my blood glucose into my target range already?

Almost. For the moment, you have received your numbers (or your FIT algorithms) for insulin dosage *without a guarantee*. They may still need to be adjusted. The point of the training program is to apply these figures and to test them under various circumstances. Together with your FIT physician, you can decide whether your algorithms have been selected accurately. If not, you can decide how they need to be changed.

I will not be so reliant on a medical team for help if I can personally adjust my own insulin algorithm, will I?

Yes, that's true. You also need to know how you can estimate whether the algorithms that you are using at that time are suitable for you, or whether they need to be changed and re-adjusted. The rules will change if your weight reduces or increases by 20 pounds. Also, your insulin requirement will change with growth hormones, menstruation, if you are ill or recovering from surgery, among other factors.

How can I tell whether I am using the right dosages for delayed acting insulin?

There are three criteria for assessing basal (fasting) insulin requirement:

1. Stability of blood glucose between meals, and over night

2. Fasting blood glucose values (ideally 90-130 mg/dl)
3. The amount of the long-acting insulin as a percentage of your total daily insulin requirement (basal/bolus balance ideally about 40%)

What percentage of the total daily insulin needs to be basal insulin?

With average nutrition, this percentage should be about 40% (50% at a maximum) of the total daily insulin requirement. Long acting insulin must always be administered in the lowest dosage possible. Delayed acting insulin is not to be used to cover food on FIT.

Long acting insulin must always be administered in the lowest dosage possible. Delayed acting insulin is not to be used to cover food on FIT.

We have already defined the fasting values. They should be 100 mg/dl in people who are not at risk of hypoglycemia. Is this correct?

It is not that easy. Can you imagine going to sleep with approximately 180 mg/dl? Where will the fasting blood glucose be then?

Equally as high or even higher?

Correct, probably even higher. The blood glucose should never decrease from the basal rate alone, at least not in type 1 diabetes.

I often have very high "fasting" blood glucose values in the morning although I go to sleep with normal values. Do I need to increase the dosage for basal insulin?

Historically, with NPH insulin for basal, high fasting values are sometimes caused by night-time hypoglycemia followed by "counter-regulation". If this is the case, you need to reduce your bedtime insulin to prevent night-time (nocturnal) hypoglycemia. With FIT, fasting values usually remain low even if you sleep through night-time hypoglycemia. It is advisable, occasionally, to measure your blood glucose when it is at its lowest (from approximately 2 a.m. to 4 a.m.) in order to ensure that you are not having hypos during the night.

And if I have excluded the possibility of nocturnal hypoglycemia?

Then you need to increase the basal rate. If using the long-acting insulin Lantus® twice daily, then "symmetrically", i.e., in the mornings and evenings. Keep in mind that the total long-acting insulin amount should not be more than half of the daily insulin dosage, at the most.

If I always have high values in the mornings despite a high basal rate, what can I do to prevent this?

Even this problem can be solved. Substituting basal insulin is sometimes difficult when the fasting insulin requirement is not distributed equally over a 24-hour period. Many people need more insulin in the morning when waking. Hyperglycemia in the morning is called the *dawn phenomenon*. This requirement for early morning insulin is defined by increased glucose production in the liver. All glucose released by the liver must be transported into the tissues by insulin. The amount of glucose production in the liver can also depend on the time of day: the highest rate is usually in the morning. It is feasible to assist the basal rate with a small amount of rapid-acting insulin since most of the day´s glucose is produced in the morning: up to 10% of the daily insulin requirement -- approximately 4 (1-6) units in most cases, can be added to the long acting insulin in the morning). This covers the increased glucose production in the morning after waking. Recently, it has been observed that even getting up in the morning causes an increase in blood glucose. Termed the *getting-up phenomenon,* this seems to accentuate the dawn phenomenon and lead to particularly high blood glucose values around breakfast time.

Using short-acting insulin (either regular or rapid-acting analogs) along with the delayed acting insulin in the morning only solves the problem of the increased insulin requirement at breakfast. What if I wake up high, have already half of the insulin allotted to the basal, and the high value is not due to a night-time hypo or skipped correction before bed time? What can I do when I still repeatedly wake up with high blood glucose values?

In this case, you need to use insulin that has its strongest effect in the early morning hours. This characteristics is found in the NPH or the insulin analog Levemir® (Fig. 4.2). These need to be injected late at night just before going to sleep, between 10 and 12 p.m. Simultaneously, it is often wise to use the analog Lantus® as morning basal insulin. This combination gives lower morning values, because both insulin types are still acting at dawn. We will come back to this later.

Instead of injecting regular or rapid-acting analogs insulin in the morning as part of the morning basal insulin, is it better to use varying insulin dosages for 1 carbohydrate choice at various times of the day? For example, in the morning, could I use 2 insulin units per 1 carb, at lunch 1 insulin unit, in the evening 1.5 units, etc.?

No, because then you will need to have breakfast to cover the *additional* requirements of the morning insulin. You can make your life easier when you compensate the varying insulin requirements throughout the day with the *basal* insulin. Keep it simple.

Is there a particular test to establish and validate the basal insulin dosage?

The only successful way to do this is to fast for one day.

It's logical to test the fasting insulin when fasting. Does that mean that I should only inject delayed acting insulin when fasting?

Yes. We will discuss this later. At this phase, you need to understand that basal means "base" or fasting insulin. Short-term fasting (between meals) should not induce a decrease or increase in blood glucose. Blood glucose should remain stable.

What should I consider for dosages of rapid-acting insulin when eating? Assuming that I know how much insulin I need for each carbohydrate choice, the dosage for a particular meal must be really easy to determine...right?

The insulin production in a person without diabetes is relatively difficult to emulate when eating. This is because a lot of insulin is produced and then rapidly surged into the bloodstream to avoid an increase in blood glucose relative to carbohydrate consumption. In contrast, with subcutaneous insulin administration in someone with diabetes, such a high insulin concentration in the blood is seldom attained. As you might know from your own experience, normal blood glucose is difficult to achieve after eating, particularly when you have eaten a lot of rapidly absorbable carbohydrate foods.

Can the normal effects of insulin be accelerated to achieve similar insulin appearance as in a non-diabetic person?

Oh, yes. By the way, the historical methods for accelerating regular insulin uptake are no more necessary when using insulin rapid-acting analogs. In general, to achieve a faster uptake of regular insulin, you can:

1. Inject regular insulin into the muscle instead of into the fatty subcutaneous tissues. The insulin enters the bloodstream more quickly since the muscle tissue has better blood circulation than fat.
2. Improve the blood circulation at the injection site by:
    - warmth
    - massaging the injection site
    - use external, circulatory-promoting anti rheumatic creams
4. Perform physical exercise.

Intra-muscular injections with regular insulin have proven to be most successful. If you inject vertically, you can easily reach a muscle in the forearm or calf and promote blood circulation at the injection site by following the guidelines mentioned in point 2. According to current research consensus, if you decide to choose insulin intravenously for exceptional reasons, the insulin dose must not exceed 10% of the total daily dosage. With this amount you will reach "maximal" sensible insulin concentration as in healthy people.

Are there any other ways in which I can achieve optimal glycemic control after eating?

As already mentioned very fast-acting insulin analogs have been manufactured by slightly changing the molecules of insulin (see Fig. 4.2.). It is much easier to achieve

lower blood glucose values after eating (even when the insulin is administered virtually simultaneously with food) when using rapid-acting analogs.
If you prefer regular insulin you can also increase the interval between injecting and eating. Particularly, before breakfast, due to *getting-up phenomenon*, 30 or even 45 minutes would be an excellent interval as long as the blood glucose is not falling to hypoglycemic levels. In this situation, we want to emphasize that accelerating the insulin absorption on a day-to-day basis is much "safer" than increasing the interval between injecting and eating. Apart from that, "long delays" before eating are not practical. Rapid-acting analogs are more effective than regular insulin for exactly these reasons.

In summary, advantages of rapid acting analogs are:

- Lower postprandial values
- Fewer intervals between bolus and meals necessary
- Quicker correction of high values
- Less hypoglycemia risks

If I understand correctly, every time I inject regular insulin I need to consider two completely separate things. First, I need to decide the insulin dosage and then I need to consider potential ways to speed up the effect of regular insulin or I must consider the interval between injecting and eating?

That's correct. You can observe this in more detail from the examples in Fig. 5.2.a/b.

So, is it true that I can only estimate the accuracy of an insulin dosage for a particular meal once the rapid-acting insulin and carbohydrates consumed have already been fully absorbed?

That is correct. Only the blood glucose values several hours (4-5h) after eating enable you to reach final conclusions regarding the selected insulin dose. Contrary to this, the value shortly after eating (1-2 hours) tells you if you have chosen the right insulin concentration for a particular meal at that point in time, i.e., whether you have correctly adapted the interval between injecting and/or the regular insulin absorption rate according to the meal.

What do you mean by "according to the meal"? Does this mean that variable insulin absorption rates (kinetics) and/or injection interval are also necessary for various types of carbohydrates?

Unfortunately, yes. The insulin requirement depends partially on the type of carbohydrate you consume and its rate of absorption (e.g. juice vs. whole-wheat bread). In reality, instead of changing the insulin kinetics, you may also reduce the absorption speed of carbohydrates by:

1. selecting more complex carbs like whole grains over refined flour products
2. increasing the proportion of non-carbohydrate sources in the meal; fat, in particular, reduces the speed at which carbohydrates are absorbed in the small intestine. You may need to be aware of the increased calories though.
3. increasing fiber with salads, vegetables, bran and guar etc.
4. drinking fewer liquids during a meal.

INTERNATIONAL RESEARCH GROUP ON FUNCTIONAL INSULIN TREATMENT
Medical University Vienna
kinga.howorka@meduniwien.ac.at
www.diabetesFIT.org

Name :..
Birth date:......................Phone:............
Address:...
E-mail..
Diabetes.since:..................Wt.:..............
FIT since:............ with O injections O pump

I BASAL (fasting insulin): AM ..../.........................U N PM ....U	**Target for correction of aberrant BG values:** Fasting/pre-meal: 100 mg/dl (or ....)
S PRANDIAL: for 1 carb choice (15g) = ...U U	After meals: 1h<180 (or.............), 2h< 140 mg/dl **Target range for MBG: from .... to .... mg/dl**
L CORRECTION: 1 U rapid ins. lowers my BG by approx.- ....; 1 carb raises my BG by approx.+ ....mg/dl	
I EXAMPLE: carb choices/CHO: .... N insulin (U):	

	TIME	1	2	3	4	5	6 AM	7	8	9	10	11	12	1	2	3	4	5	6	7 PM	8	9	10	11	12	Total Daily Dose
MON	Basal																									
	Bolus	Regular insulin										5														
7.8.	BG												100			220			104							MBG
	Carbs/CHO													3												
	Comments	Example 1																								
TUE	Basal																									
	Bolus												4													
8.8.	BG												90			146		96								MBG
	Carbs/CHO													3												
	Comments	Example 2																								
WED	Basal																									
	Bolus													7												
	BG											110			120		~40									MBG
	Carbs/CHO													3												
	Comments	Example 3																H								
	TIME	1	2	3	4	5	6	7	8	9	10	11	12	1	2	3	4	5	6	7	8	9	10	11	12	Total Daily Dose
THU	Basal																									
	Bolus													3												
	BG											95			140			190								MBG
	Carbs/CHO													3												
	Comments	Example 4																								
FRI	Basal																									
	Bolus																									

Fig. 5.2.a: Meal related insulin

You can see here four situations where regular insulin was injected when eating. Try to answer the two following questions:

1. Was the insulin dosage chosen correctly?
2. Was the timing of the injection well-chosen according to the meal, was the correct "speed of insulin absorption" achieved (physicians call this "insulin kinetics")?

In all these examples, we assume that the dosage for basal insulin is correct, i.e. no blood glucose fluctuations are due to basal insulin dosage.

Fig. 5.2.b:

Example 1:
One to two hours after eating, the blood glucose level increases strongly! The carbohydrates consumed were absorbed more quickly than the insulin administered. If a higher dosage had been chosen, blood glucose values could have been lower after eating, but hypoglycemia would result later.
Conclusion: right dosage, wrong "insulin kinetics".
To avoid the sharp increase of blood glucose after eating, the following precautions could be taken:
1. Lengthen the inject-eat interval
2. Speed up the regular insulin absorption or
3. Use rapid-acting analogs

Example 2:
In this case, the blood glucose values are within the target area, approximately 1.5 hours after eating and even 5 hours after eating.
Conclusion: right insulin dosage and right method of insulin administration (in this case, the inject-eat interval was longer).

Example 3:
The blood glucose values after eating are excellent: 120 mg/dl. However, this does not mean that the dosage was correctly chosen. Several hours after eating, this leads to hypoglycemia, which indicates that the chosen insulin dosage for the meal consumed was too high.
Conclusion: Short-acting insulin dose was too high.

Example 4:
The blood glucose values are still in the target level immediately after eating. However, several hours after eating, after the complete absorption of carbohydrates and insulin, the blood glucose is 190 mg/dl, above the target area. Assuming a correct dosage of basal insulin was provided, we conclude that the insulin dosage for lunch was too low.
Conclusion: Regular insulin dosage was too low.

Is there a test to find out if I correctly injected the insulin for a meal?

Yes, since you already have experience with rapid-acting insulin, and when you have adjusted your algorithms to "so many units of insulin per one carbohydrate choice", you are now able to "sin" a little. The test is called a “celebration day”, “yielding to temptation”, or simply “a sin”. Simply eat something that you have wanted for a long time but did not risk eating before. From the blood glucose response, you can see immediately whether you:

1. have correctly estimated the quantity of carbohydrates in the food that you chose
2. have chosen the appropriate insulin dosage for it
3. have correctly chosen the absorption rate or the inject-eat interval
4. can carry out all of these tasks on your own even outside the clinical setting, without scales or other hospital equipment.

It seems to me that I’m becoming a specialist in "celebrations"...

You should try the experiment of the so-called *"celebration day"* when you have obtained experience with FIT, and above all, when you clear about your algorithms of insulin dosage.

How important is it to consider the non-carbohydrate sources in a meal? Up to now, I have been told that I can eat foods high in protein and fat without injecting insulin or having high blood glucose.

If you eat a low-carbohydrate meal (carbohydrate share: less than 30%) that is high in protein, your liver will eventually produce more glucose or carbohydrate units. If you’re trying to avoid an injection of rapid-acting insulin, eating a sausage instead of a bun is not the answer, because you will still need some short-acting insulin, though much less, of course, to cover meals **low in carbohydrates**. With a usual daily insulin requirement of 40-50 units, approximately half a unit of short-acting insulin is required per 100 calories of a protein-fat mixture in such meals.

Adding protein and fat to a meal where the proportion of carbohydrate is 40-50% will slow down the intestinal absorption of carbohydrates and even somewhat reduce the insulin requirement for this meal. The quicker the absorption of carbohydrates, the higher the insulin requirement.

What do I need to consider when correcting overly high blood glucose?

Be clever and don’t eat while your blood glucose is high. This is not an exaggeration. Increased blood glucose correlates with the lack of insulin in type 1 diabetes. If you eat without an insulin supply or even immediately after correcting overly high blood glucose, you can expect blood glucose to increase greatly. Therefore, after the correction wait before eating until your blood glucose is in the target area. Eating carbohydrates when there is not enough insulin it is not advisable because the excess glucose will be discharged through the kidneys as urinary glucose.

Do I need to note anything else if I want to lower my blood glucose?

Yes, you need to beware of doing what is known as "double corrections". Allow your blood glucose time to decrease. The shortest interval between two consecutive blood glucose corrections should not be less than 2-3 hours with rapid-acting analogs or 3-4 hours with regular. During this time, after a blood glucose correction, use rapid-acting insulin only to cover food (prandial needs) but not to correct high blood glucose (hyperglycemia). Otherwise, an insulin overdose and hypoglycemia can result.

During 2-3h after the meal, rapid-acting insulin is injected to cover food only (prandial needs), not to correct high blood glucose (hyperglycemia)

So do I understand that measuring blood more frequently than every 2-3 hours is not very sensible?

That's right. More frequent blood glucose measurements at intervals of less than 2-3 hours are not useful because they do not allow any insulin dosage consequences. However, during the learning phase of FIT you will be measuring more often. Only by doing so can you develop a "feel" or awareness for various factors that influence your blood glucose. This will save time and effort for you in the future. Your treatment will be more effective in the long run.

Is there a test to validate algorithms for blood glucose corrections? I really want to determine by how much (expressed in mg/dl) 1 unit of rapid-acting insulin lowers my blood glucose and by how much 1 carbohydrate choice raises my blood glucose.

You can try, under controlled circumstances, to increase your blood glucose with glucose tablets or lower it with rapid-acting insulin after you have adhered to your rules for several days. If you test urinary and blood glucose values at the same time, you may determine your "kidney threshold". The "threshold" in this case is any blood glucose value in which the urinary glucose is positive. We will return to this in the chapter about insulin games.

So, what is most important when using delayed acting insulin?

The delayed acting insulin is used as a basal (or fasting) insulin under FIT conditions. Basal insulin can also be replaced by means of continuous infusion of rapid-acting insulin using an insulin pump.

To assess the accuracy of the basal dosage, the following criteria need to be considered routinely:

1. The stability of blood glucose values between meals. If the basal insulin dose is correct during short-term fasting (up to maybe 14 hours, e.g. at night-time) there will not be any spontaneous decrease or increase in blood glucose

(except if induced by physical exercise).

2. Presuming a late evening blood glucose correction targeted at approximately 120 mg/dl, we expect that the fasting blood glucose values (the following morning) would only rarely exceed 90 - 140 mg/dl.
3. The ratio of delayed acting insulin to rapid-acting insulin (basal bolus ratio). At the most, 50% of the daily insulin requirement should be allotted to delayed-acting insulin; optimum is usually approximately 40%.

If, despite a relatively "high" basal rate, fasting blood glucose values are regularly high (dawn phenomenon), then the basal rate would need to be corrected (providing nocturnal hypoglycemia was excluded).

Better fasting values can sometimes be reached with Lantus® used once daily before dinner. Alternatively, insulin of the NPH-type or Levemir® is to be used late at night before going to sleep, with Lantus® in the morning for much better fasting values. On the contrary, fasting blood glucose values under 90 mg/dl often indicate that basal insulin is too high.

In the morning, a small amount of rapid-acting insulin often needs to be included in the basal rate, even if you do not eat breakfast! This prevents hyperglycemia later on in the morning ("morning mound" for *getting-up phenomenon*).

What is most important when using short-acting insulin?

Short-acting insulin is used under FIT conditions for the following purposes:

1. To lower a high blood glucose value i.e. for *correctional* purposes;
2. To serve as *a meal-related* insulin replacement i.e. for prandial purposes
3. To balance out the increased requirements for insulin in the morning ("morning mound" for getting-up phenomenon).

In all three situations, insulin is lacking and is needed as soon as possible. To exaggerate somewhat -- if a short-acting insulin is necessary at all with FIT then it will be needed quite urgently! As subcutaneously injected regular insulin enters the bloodstream quite slowly, certain measures can routinely be taken to speed this up or modify "insulin kinetics". Except for those people who have delayed emptying of the stomach due to neuropathy, rapid-acting analog is a good alternative to regular insulin.

When using rapid-acting insulin *at meal-times*, the following facts need to be taken into consideration (assuming correct basal insulin):

1. Only late postprandial values 4-5 hours after eating give definite information about the appropriateness of the chosen dose of short-acting insulin.
2. Early values 1-2 hours after eating on the other hand, give information about the injection-eat interval as well as the chosen insulin absorption rate. Final conclusions regarding the chosen dose of insulin are only possible when both, prandial insulin and the food are completely absorbed through the intestines, 4-6 hours later.

When using rapid-acting insulin *for correcting hyperglycemia* (high blood glucose), the following rules are recommended:

1. If possible, eat only when blood glucose is back to your target range (i.e. after lowering a high blood glucose level).
2. Double corrections need to be avoided, i.e. after correcting hyperglycemia with rapid-acting insulin, do not use further correctional insulin for at least 3 hours. Insulin can be injected for prandial (meal-related) needs within shorter time intervals.
3. The current target for blood glucose correction always needs to be taken into account. One to two hours after eating, blood glucose values are acceptable up to 160 mg/dl, even up to 200 mg/dl in people who are at particularly high risk of hypoglycemia.

You can find the summary of the Functional Insulin Treatment program, with short tests and daily criteria for evaluating FIT algorithms on the next page. Included are also simple equations deriving initial algorithms for FIT from your total daily dose and your body weight. Check with your physician whether the initial algorithms derived as indicated can be used in your particular situation.

<table>
<tr><td colspan="3">Mathematical equations for initial algorithms are derived from our nomogram for initial FIT-algorithms (K. Howorka, Functional Insulin Treatment, Springer Publishers)
ATTENTION: Please ask your physician after the estimation of your algorithms whether these algorithms can apply for your particular case!</td></tr>
<tr><th>Algorithms of FIT</th><th>Short tests: “Insulin games”</th><th>Assessment criteria for every day</th></tr>
<tr><td>Basal insulin
How much (and which) insulin do I need even if I don’t eat anything?
Total Daily Dose (TDD) x 0.45 = ____________
(of which 10-20% is used as “morning mound” + 80-90% as delayed acting insulin)</td><td>Basal insulin
One-day-fasting (should require maximum 2-3 carbohydrate choices to keep up with basal rate without prandial insulin when fasting for 36 hours)</td><td>Basal insulin
• blood glucose stability with short-term fasting (between meals)
• fasting values (mostly between 90-140 mg/dl)
• daily proportion (balance) delayed-acting insulin to short-acting insulin</td></tr>
<tr><td>Prandial insulin
How much short-acting insulin do I need for 1 carbohydrate choice?
TDD x 0.04 = __________
(average insulin need per 1 carb choice)</td><td>Prandial insulin
So-called „yielding to temptation“ or “celebration day” (correct if blood glucose in the target earlier and later after the meal)</td><td>Prandial insulin
• blood glucose, short-term (1-2 h) after eating: information mostly about absorption rate of insulin and carbohydrates. Prandial insulin dosage still can not be eventually determined
• only blood glucose late (4-6 h) after the meal allows reliable judgment of the insulin quantity used for this meal</td></tr>
<tr><td>Blood glucose correction algorithms
By how much mg/dl does 1U of short-acting insulin lower my blood glucose?
1700 : TDD = __________
BG lowering by 1 U insulin
By how much mg/dl does 1 carbohydrate choice increase my blood glucose?
90 – (body weight in kg x0,5) = BG increase by 1 carb</td><td>Blood glucose correction algorithms
Checking the correction algorithms and determining the kidney threshold for glucose, as a “side-effect”</td><td>Blood glucose correction algorithms
Judging of correction insulin dose only after completion of absorption, see above; individual risk of hypoglycemia determines target of blood glucose</td></tr>
</table>

Fig 5.1.: Short test and all-day criteria for assessment of algorithms for insulin dosage for FIT. Note:

TDD = Total Daily Dose, with usual eating habits and acceptable glycemic control (MBG=150-250 mg/dl). Calculation of BG lowering by 1U insulin applies *only* if proportion of basal insulin is below 40-50% of TDD! A transfer to pump treatment will often lower total daily dose by 10-20%.

Test Questions on Phase Two: FIT Training

Name: .... Date:....................................

A) Decide whether the following sentences are true or false:

		true	*false*
1.	Anyone can learn FIT in two days after reading this book (FIT patient manual)	□	□
2.	Algorithms of FIT describe how to dose insulin for fasting, eating or correcting blood glucose.	□	□
3.	The participation in FIT training is reasonable after a comprehensive, structured general diabetes education.	□	□
4.	It is sensible to begin FIT after self-monitoring several times a day.	□	□
5.	FIT allows you to correct your blood glucose, but you must never change your algorithms by yourself.	□	□
6.	FIT is designed only for people who are able to take complete responsibility for their own treatment.	□	□
7.	Treatment allows good metabolic control but never a flexible diet.	□	□
8.	It is not important to know how much insulin you need for 1 carbohydrate choice to start FIT.	□	□
9.	The basal fasting insulin requirement is the same for all people with diabetes.	□	□
10.	General basic diabetes training and special FIT training may be completed within 4 days.	□	□
11.	"Insulin games" are included in FIT training to determine if you can manage without insulin.	□	□
12.	Personal algorithms of insulin dosage remain the same for every person for their whole life.	□	□
13.	FIT frees people with diabetes from blood glucose self-monitoring.	□	□
14.	The most important blood glucose measurement in FIT is the one before going to sleep.	□	□
15.	You need a calculator for FIT to calculate how much insulin you need for meals.	□	□
16.	Even large quantities of protein and fat may be eaten under FIT without insulin.	□	□
17.	FIT is the abbreviation for Functional Insulin Treatment.	□	□

B) A person with type 1 diabetes on FIT can do without any insulin by eating only the following:

		true	*false*
1.	A sausage, roll and small beer	□	□
2.	A large pork chop and cucumber salad with dressing	□	□
3.	A large green salad with lemon juice	□	□

C) During Phase Two of structured FIT group education you need to carry out the following self-monitoring (check all that apply):

		true	*false*
1.	blood glucose twice daily	□	□
2.	blood glucose 4 times daily	□	□
3.	blood glucose 8-10 times daily	□	□
4.	urinary glucose and ketones several times a day	□	□
5.	urinary glucose and ketones not at all	□	□

D) Imagine you are a person with type 1 diabetes and the following rules (algorithms) of insulin dosage with FIT:

* **Basal insulin:**
Mornings: 12 U Lantus® + 4 U rapid-acting insulin
Evenings: 12 U Lantus®

* **Prandial insulin:**
1.5 U rapid-acting insulin for every carbohydrate choice:

* **Correction algorithms:**
1 U of rapid-acting insulin lowers my blood glucose by - 40 mg/dl
One carbohydrate choice increases my blood glucose by + 50 mg/dl
Target blood glucose levels:
fasting and before eating: 100 mg/dl
1-2 hour after eating: up to 180 mg/dl

What would you do in the following situations? Decide whether the responses given are true or false.

1. You wake up at 7 a.m. and your blood glucose is 180 mg/dl. You would like to eat a buttered medium-sized roll and 1 cup of yogurt for breakfast. You inject the following:

		true	*false*
a)	11 U rapid-acting insulin, 12 U delayed acting insulin	□	□
b)	12 U delayed acting insulin, 6 U rapid-acting insulin	□	□
c)	7 U rapid-acting insulin, 14 U delayed acting insulin	□	□

2. It is 1 p.m. and several hours have passed since eating and injecting. You measure a blood glucose of 240 mg/dl. You don't want to eat anything. Lunch will obviously be postponed today as you have to go to an urgent meeting with your boss. You inject:

		true	*false*
a)	3 U delayed acting insulin, 0 U rapid-acting insulin	☐	☐
b)	3 U rapid-acting insulin, 0 U delayed acting insulin	☐	☐
c)	5 U rapid-acting insulin, 0 U delayed acting insulin	☐	☐

E) Imagine you have type 1 diabetes and the following rules or algorithms of insulin dosage on FIT:

* **Basal insulin:**
  Mornings: 18 U Lantus® + 6 U rapid-acting insulin
  Late evenings: 17 U Levemir®
* **Prandial insulin:**
  2 U rapid-acting insulin for every carbohydrate unit
* **Correcting values:**
  1 U of rapid-acting insulin lowers my blood glucose by - 35 mg/dl
  One carbohydrate choice increases my blood glucose by +60 mg/dl
  Target blood glucose levels:
    fasting/before eating 100 mg/dl
    1 hour after eating up to 180 mg/dl

What would you do in the following situations?

1. You wake up with 120 mg/dl. It is already 7:20 a.m. You have no time for breakfast. You inject:

		true	*false*
a)	17 U Levemir®, 6 U rapid-acting insulin	☐	☐
b)	18 U Lantus®, 0 U rapid-acting insulin	☐	☐
c)	6 U rapid-acting insulin, 18 U Lantus®	☐	☐

2. It is 11 p.m. You want to go to sleep. Several hours have passed since eating dinner. Your blood glucose is 190 mg/dl. You inject:

		true	*false*
a)	18 U Lantus® insulin, 4 U rapid-acting insulin	☐	☐
b)	2 U Levemir®, 17 U rapid-acting insulin	☐	☐
c)	2 U rapid-acting insulin, 17 U Levemir®	☐	☐

F) Imagine you are a person with type 1 diabetes and the following rules or algorithms of insulin dosage on Functional Insulin Treatment:

* **Basal insulin:**
  Mornings: 7 U Lantus® + 2 U rapid-acting insulin
  Evenings: 7 U Lantus®
* **Prandial insulin:**
  1 U rapid-acting insulin for every carbohydrate choice
* **Correcting values:**
  1 U of rapid-acting insulin lowers my blood glucose by - 50 mg/dl
  One carbohydrate choice increases my blood glucose by +50 mg/dl

Target blood glucose levels:
fasting/before eating: 120 mg/dl
2 hours after eating: up to 200 mg/dl

What would you do in the following situations? Mark true or false.

1. It is 7 a.m. You wake up with blood glucose of 60 mg/dl. You would like to eat oat meal (2 carbohydrate units). You inject:

		true	*false*
a)	5 U rapid-acting insulin, 7 U Lantus®	☐	☐
b)	3 U rapid-acting insulin, 7 U Lantus®	☐	☐
c)	0 U rapid-acting insulin, 3 U Lantus®	☐	☐

2. At 1 p.m. you measure blood glucose of 220 mg/dl. Several hours have passed since eating and injecting. You would like to eat 3 carbohydrate choices at lunch time. You inject:

		true	*false*
a)	3 U rapid-acting insulin, 7 U Lantus®	☐	☐
b)	5 U rapid-acting insulin, 0 U delayed acting insulin	☐	☐
c)	7 U rapid-acting insulin, 0 U Lantus®	☐	☐

6. Summary of the First FIT Day

One hour after breakfast, my blood glucose was 260 mg/dl!

What was the blood glucose when fasting?

Not that good either. It was 150 mg/dl. Did I correct too little then?

Three factors define what the rapid-acting insulin dosage should be in the morning before breakfast:

1. Correction of the blood glucose, if it is beyond the target point.
2. The balance of the basal insulin requirements in the morning with the prescribed quantity of rapid-acting insulin (about 4 units in adults). This rapid-acting insulin dosage is also known as "*morning mound*". You need it for the increased glucose production, which takes place in the liver early in the morning.
3. Prandial insulin – if you want to eat breakfast.

I did all of that. Injected the correction insulin,* my morning mound *and prandial insulin. Despite carrying out all three steps, my blood glucose increased after breakfast. Should I have taken more insulin for my* morning mound*?

You can barely see whether the insulin dosage was right or not from the value. 1-2 hours after breakfast. You can judge this only 4-5 hours after eating (to be more exact: after injecting the prandial insulin). In other words, only when breakfast and the insulin injected for breakfast time have been absorbed completely and have exerted their effect can you judge, by the blood glucose value, whether the chosen dosage was correct. It was correct if the blood glucose is in the target level, too high if you have hypoglycemia, or too low when the blood glucose values are still high, perhaps before eating lunch.

I did not do anything to speed up the insulin effect but I waited more than half an hour after injecting my regular insulin before starting the meal.

As you have seen, this was not enough. The intake of food when blood glucose is high can lead to even higher values. Of course you can eat anytime, if you want! Right now you are deciding independently if, what, and when you eat. However, when you eat breakfast with a blood glucose of 160 mg/dl without "speeding up" the insulin absorption, it is not surprising if the values are so high after breakfast.

Does the increase in blood glucose after eating have anything to do with mixing both types of insulin: delayed-acting and regular insulin?

You shouldn't mix Levemir® or Lantus® insulin with regular insulin or short acting insulin analogs. Even mixing with NPH changes the effect of regular insulin, but to a lesser degree than with zinc insulin (not available at present any more). It is, therefore, advisable to inject each type of insulin separately and to avoid mixing them in one syringe.

One hour after breakfast when my blood glucose was 260 mg/dl, I corrected it with 4 units of rapid-acting insulin to 100 mg/dl. The blood glucose decreased, but before lunch I had a "hypo". Did I take too much insulin for correcting?

You made several poor choices. First, you chose the wrong target point.

That's right. I should have corrected to the post meal target point below 160.

Yes, 160 or even 180 mg/dl, if you tend toward serious hypoglycemia. But, you also made another mistake. You corrected about 1 hour after eating and previously also prior to breakfast, not even 2 hours before. Remember? You had already corrected your fasting blood glucose!

So I corrected it "twice"?

Exactly!

No wonder I had a "hypo" before lunch! I'll try not to do that again, for sure. But, why were my fasting blood glucose values so high? Should I increase my basal rate?

The fasting values alone are of minor significance when judging the basal rate. You went to sleep with blood glucose of 170 mg/dl, without correcting it to your target area before sleeping. At this point, your dinner had been already completely absorbed several hours before.

So I should have corrected my blood glucose to the target area around 100 mg/dl in this case?

Not necessarily yesterday. It was good that you waived this correction on the first day because you did not know whether the suggested basal rate was right for you. In other words, you temporarily and sensibly chose a higher target level for safety reasons. Of course, it is not surprising that your blood glucose values were higher in the morning.

I measured my blood glucose at 3 a.m. The value was 150 mg/dl. If the following three values
(1) late before going to sleep,
(2) at the lowest point, about 4 a.m. and
(3) fasting
are similar, does this mean that my basal rate is correct?

Exactly! But if you started with your new treatment yesterday, it is still too early to answer this question properly. You have used other insulin since yesterday. You need to wait a couple of days before you can judge the accuracy of your basal rate. Everyday criteria for evaluating basal insulin dosage are:

(1) blood glucose stability between meals,
(2) fasting values, and
(3) the daily relationship between delayed-acting and rapid-acting insulin. We will return to this later

I had some peaks above my target level. However, during this first day, I was able to lower my average daily blood glucose to less than 140 mg/dl.

It is important not to underestimate the importance of individual peaks. It is good if you can maintain your average blood glucose values under 140 mg/dl. But, you should avoid playing "yo-yo" games with your blood glucose. To avoid the high values, do the following:

1. Correct your blood glucose. The most important corrections are late before going to sleep, fasting in the morning, and perhaps twice more during the day.
2. "Speed-up" the action of regular insulin when you have high blood glucose, eat carbohydrates, and need increased insulin in the morning, or take rapid-acting analogs for carbs or corrections.
3. Extend the injection to eating interval to reach optimal blood glucose values after breakfast.
4. "Don't eat when your blood glucose is elevated"; first lower it with short-acting insulin.

What is important for avoiding low blood glucose values and hypoglycemia?

Always try to use the smallest amount of insulin as possible. In reality, this means:

1. Always choose a safe target point for blood glucose corrections for you. Values of 160-180 mg/dl 2 hrs after eating are acceptable.
2. Blood glucose corrections after meals are possible but relatively dangerous. Avoid them as much as possible. Nonetheless, if you decide to make a correction after eating, then select a postprandial target of 160-180 mg/dl, and only if you have not made a correction in the past 2-3 hours.

Over the next few days, during transition, it is not important to have "wonderful" blood glucose values. Instead, try to interpret your blood glucose levels and search for reasons why it might be above or below the target level. If it is too high or too low, correct your blood glucose value. At the same time, recall which and when you took insulin. Was it too much or too little? What could you do in the future to avoid this? Learn to judge whether your algorithms of insulin dosage (specially suggested for you) are correct or not. Suggest how your rules of insulin dosage could be adapted to suit "you". In reality, this means that you will establish your own dosage guidelines and continue to do this in the future for yourself.

7. "Insulin Games" to Check Insulin Dosage Algorithms

You mentioned some tests for checking my insulin dosage guidelines?

You can check your prandial insulin guidelines by applying them at meal-times. You can check your fasting insulin dosage by fasting and your blood glucose correction algorithms by intentionally increasing or decreasing your blood glucose. This is usually combined with the estimation of your kidney threshold for glucose.

7.1 Meal-related Insulin. The "Celebration Day" or "Yielding to Temptation"

Is "Yielding to Temptation" a test to check the algorithms for food choices?

Not exclusively. Once you choose to eat whatever you want, you need to test your prandial insulin replacement ability under "real life" conditions.

I think I'm ready to "sin" a little!

"Yielding to Temptation" in the context of your Functional Insulin Therapy does not exist anymore, because you can eat whatever and whenever you want and still stay in very good control of your diabetes, if you are able to use insulin flexibly. Do only what you are capable of doing. In this exercise, you test your knowledge and discover your capabilities -- including their limitations -- in a protected environment during training. Remember that healthy eating is still a good habit, just like it is for everyone else. A specific record sheet for "Yielding to Temptation" is important to interpret results correctly and to improve your abilities for the future.

So, the normal blood glucose log book is not sufficient for the "Yielding to Temptation" experiment?

No, we recommend a special log book just for this purpose. You can see an example in Figure 7.1. An "empty" record sheet for this exercise you will find at the end of this book.

Can I assume that I have chosen the right quantity of insulin for my meal when my blood glucose is correct after eating?

Values shortly after meals (approx. 1-2 hours later) are important to enable you to judge whether the insulin absorption speed corresponds with the absorption rate of food eaten. But, the values at this time do *not* tell you whether the chosen insulin dosage was right.

So, in my understanding, only the "late" values after meals (about 4-6 hours afterwards) determine the correctness of the insulin dose?

That's right, after both the insulin and the food have been absorbed. If you are on target value with your blood glucose several hours after injecting and eating, you can assume that the insulin dosage for this meal was correct, as long as your basal insulin dosage is correct.

With "Yielding to Temptation" I can check also my correction algorithm if my blood glucose is not exactly on target?

That's right. "Yielding to Temptation" supports your ability to register and correct your current blood glucose on a day-to-day basis with your minimal equipment, and to dose insulin correctly.

Experiment: 'Celebration Day' or 'The Yielding to Temptation'

Name: ***Caroline***
Date:

Eat what you want.

Whilst you are doing this (and afterwards!) answer the following questions:

1. *Insulin dose:* Can I choose the correct insulin dosage for a given meal?

   Yes!

2. *Insulin kinetics:* Can I choose the most suitable mode of insulin administration and the appropriate interval between injecting and eating?

   Not yet...

3. *Practical skills:* Have I got everything I need in my "mini-kit"? That is, am I equipped to correctly estimate my blood sugar (even perhaps without a glucometer) and to administer insulin when I am on my own?

   Yes!

Time	Blood glucose [mg/dl]	Insulin [units]	Carbohydrate choices / meal	Notes
14:30	*194*	*2 units Regular (correction)*		*We are going to St. Stephen's Cathedral*
15:45	*82*			*Tea in coffee house*
		7 units Regular		*Ordered apple strudel*
16:10			*3-4 Carbs*	*Apple strudel*
16:30	*242*	*No correction*		
17:30				
18:00	*121*			*Back to Kärtner street on foot*
18:45				*Injected at evening meal*
4:00	*110*			
7:00	*126*			

Fig. 7.1: "The Yielding to Temptation"

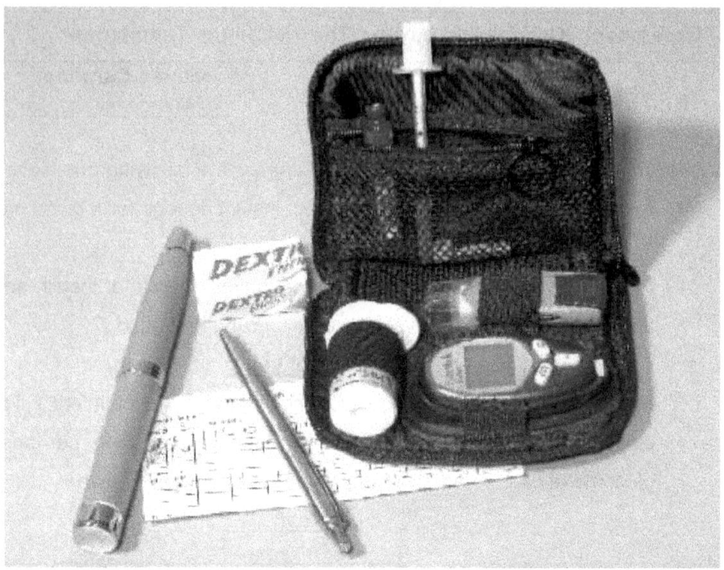

Fig. 7.2: The minimum equipment for FIT. Here an example with an insulin pen.

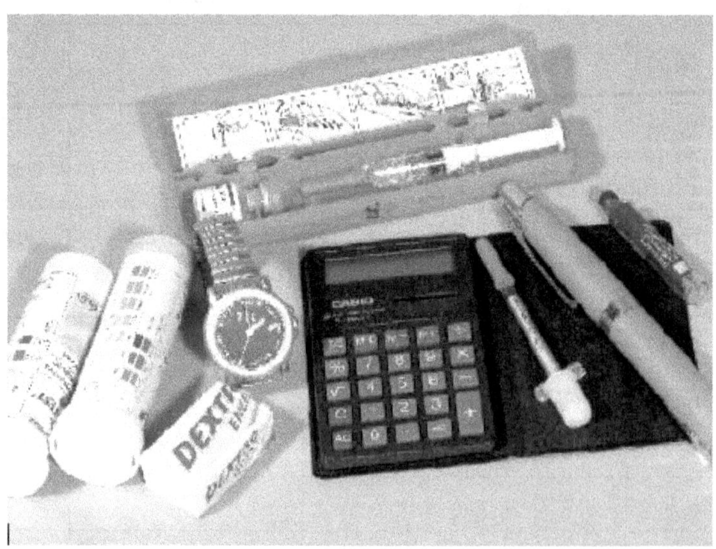

Fig. 7.3: The "extended" minimum equipment for FIT: test strips for urinary ketones, visual blood glucose test strips (see: www.betachek.com) and a watch with seconds display, calculator for calculating average blood glucose, delayed acting insulin, insulin syringes (even if you use only an insulin pen) and glucagon injection kit.

What is the "minimal equipment" required for FIT in everyday life?

The following is indispensable (Figure 7.2, mini-kit for FIT)

1. Small, efficient blood glucose meters (alternatively, if available, visual blood glucose strips that test for blood glucose, e.g. Betachek visual strips, see also www.betachek.com),
2. Rapid acting insulin (as you only inject delayed-acting insulin once or twice a day, you do not have to take it with you as long as you return home at night),
3. An insulin syringe or a pen-injector,
4. Glucose tablets,
5. Your log book and pen.

The items mentioned are *essential.* They should be light and easy to carry. Otherwise you will not be able to carry them with you all the time. Many people prefer an insulin pen to a syringe.

How often do I need to measure blood glucose when I "Yield to Temptation" outside of a diabetes training center?

At this point, you are still in a learning phase. Measurements should be taken more often at first, then later on necessary. This way you can discover limitations of your abilities. If your blood glucose was between 80 and 160 mg/dl you are really an expert. If it was between 60 and 240 mg/dl when "Yielding to Temptation" your knowledge and capabilities are "acceptable". If your blood glucose was over 240 or under 60, then you need to reconsider. Try to analyze (with a professional FIT adviser) what you need to do to improve your situation in the future.

Is "Yielding to Temptation" validated and completed by measuring blood glucose 4-5 hours after eating?

Not quite. The effects of food intake and rapid insulin can occur even much later. If you had 80-120mg/dl blood glucose during or after "Yielding to Temptation," but experienced hypoglycemia 6-7 hours later at about 3 a.m., your "Yielding to Temptation" was not really perfectly managed. Or, if you eat more than usual on a particular day as an exception to a certain extent you will also "stuff" your liver's storage with excess glucose. It is well known that increased food intake leads to increased blood glucose and an increased "dawn" phenomenon the next morning. You are a real *"FIT-pro"* when you are able to manage the potential longer term consequences of "Yielding to Temptation" and also maintain acceptable blood glucose values the next morning.

Any special solutions for "Yielding to Temptation" with an insulin pump?

Yes, please take into account that some meals will require a delayed or combined ("dual") bolus (e.g. for fatty American pizza with a lot of cheese). Only small and carbohydrate rich meals can be covered with standard bolus.

How relevant is the exact time of the day for the prandial dosage requirement?

As already discussed, for dawn and dusk you will need some more insulin. I recommend compensating these circadian rhythms with basal insulin throughout the day. It is more practical, and you can use the same rule for the prandial insulin (insulin to carb ratio).

Can you summarize?

Yes. Correct choice of prandial insulin dose can be recognized on near-normal blood glucose values several hours after eating. Short-term postprandial values are an indicator for prandial insulin absorption and not necessarily an eventual proof for the prandial dose used.

7.2 Correction Insulin. Checking Blood Glucose Correction Algorithms. "Side Effect": Determining Your Kidney Threshold

How does the determining of kidney threshold for glucose relate to FIT algorithms?

You can investigate the influence of carbohydrates and then short-acting insulin on blood glucose in *your* body by intentionally increasing and decreasing blood glucose. In doing so, you need to use *your* correctional algorithms. You will additionally gain more information on your body if you combine this experiment with determination of the kidney threshold for urinary glucose. As already mentioned, kidney threshold for glucose indicates blood glucose values, at which, when reached, the kidney begins to remove glucose into the urine.

How do I actually determine my kidney threshold for glucose?

The kidney threshold test needs to be started with a near normal (i.e. below the kidney threshold) blood glucose. No glucose should be excreted into the urine. You can intentionally raise blood sugar by eating so many glucose tablets that the blood glucose value reaches about 240 mg/dl. You test your blood sugar and urinary glucose at frequent intervals during this intentional blood glucose increase. In doing this, you will discover the interdependence between blood glucose values and sugar in the urine – "the kidney threshold for glucose". You will need to drink a lot of sugar-free beverages such as water, tea or diet Pepsi/Coke, to produce a lot of urine during the testing period. You need to test your blood and urine for glucose every ten minutes or so. Wait until all of the glucose consumed is distributed throughout your body before you decrease your blood glucose to the target area with insulin.

Wait until your blood glucose is stable and is no longer spontaneously decreasing. Now, you can intentionally lower your blood glucose by calculating how many insulin units you need to lower your blood glucose to the target. At the same time, you can check the algorithms used until now: "1 unit of rapid insulin lowers my blood glucose by ___ mg/dl". Figure 7.4 a/b shows an example of a record sheet used to determine the kidney threshold for glucose.

It looks like the whole check-up takes a long time from this record sheet.

Unfortunately, it does. The longest and most difficult phase is reaching "basal" conditions.

What does "basal conditions" actually mean?

You will reach the basal conditions when your blood glucose is stable without large fluctuations. In other words, "fasting" insulin currently in your blood closely matches the quantity of glucose released by your liver. Only the amount of glucose produced is transported into your cells. You'll have s*table* blood glucose without eating.

Now I understand why the "kidney threshold game" is recommended only several hours after a meal and after an injection of short-acting insulin.

This is necessary because otherwise it is difficult to interpret the course of the blood glucose level.

Is it really so difficult to keep the blood glucose stable?

You need to try to keep it stable. You will soon see that -- despite your best efforts -- your blood glucose often has a certain spontaneous dynamics of its own. Generally, you will have either too much insulin causing a spontaneous blood glucose drop or too little insulin causing an increase in blood glucose exceeding the kidney threshold.

Patient: ***Susan M.***	Date: *05.03.*	Trainer: *Dr. Grill.*

Technical requirements Fulfilled: **Yes** **No**

		Yes	No
1.	Glucose meter with adequate strips available?	☑	☐
2.	Ketones strips available?	☑	☐
3.	Urine glucose strips available?	☑	☐
4.	Sufficient fluid available (approx. 2litres,but no beer or milk)?	☑	☐

In order to start the test, you need to be able to answer "yes" to all questions. If not, please carry out the test another time when all technical requirements can be fulfilled.

Phase 1: Fulfilling the prerequisites for kidney threshold determination. Achieving basal conditions.

		Yes	No
1.	Last meal at least 4 hours ago? ***(4,5h)***	☑	☐
2.	Last rapid acting insulin injection at least 4 hours ago?	☑	☐
3.	Blood sugar stable for at least 1 ½ hours? (No upward or downward trend)	☑	☐
4.	Urine sugar negative? ***(traces at midday)***	☑	☐
5.	Urine ketones negative?	☑	☐

If you have answered "no" to one or more of these questions, the kidney threshold test will not give meaningful results. In this case you need to conduct the test on another day when you are able to achieve basal conditions. During Phase 1 you need to drink at least 1 liter (four 8 oz. glasses) of any calorie-free liquid (water, mineral water, tea, etc.)

Phase 2: Controlled raising of blood glucose (Questions 1 and 2).
Eat enough glucose to raise your blood glucose to about 250 mg/dl. Try to pass enough urine to measure urine sugar in 10-min intervals. (This is relatively easy if you have consumed at least 1 liter of fluid during Phase 1). Measure your blood sugar at the same time that you measure your urine sugar. Drink another ¼ to ½ liters (one to two 8 oz. glasses).

Phase 3: 'BG-Plateau'.
Wait until your blood sugar has stabilized and does not drop spontaneously. Only then can you use your algorithms to calculate the amount of rapid-acting insulin you need to lower your blood sugar to about 100-110 mg/dl. Drink another ¼ to ½ liters (1–2 8 oz. glasses).

Phase 4: Controlled lowering of blood sugar with regular insulin.
Inject the amount of rapid acting insulin you have calculated that will be necessary to lower your blood sugar to 100-110 mg/dl. In order to save time, use the methods you have learned to speed up the action of rapid-acting insulin. Answer Question 3. Continue to measure both urine and blood sugar in 15 minute intervals.

Fig. 7.4.a: Worksheet for determining the kidney threshold

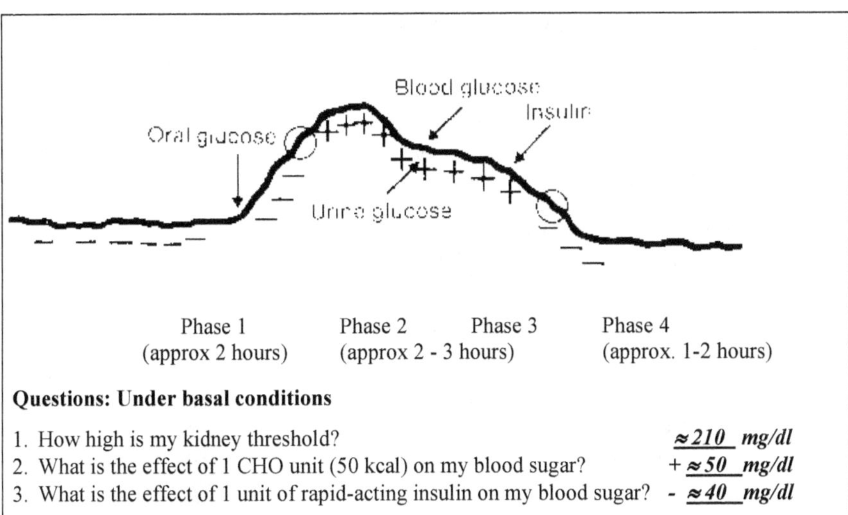

Phase 1 (approx 2 hours) | Phase 2 (approx 2 - 3 hours) | Phase 3 | Phase 4 (approx. 1-2 hours)

Questions: Under basal conditions

1. How high is my kidney threshold? ***≈210* mg/dl**
2. What is the effect of 1 CHO unit (50 kcal) on my blood sugar? + ***≈50* mg/dl**
3. What is the effect of 1 unit of rapid-acting insulin on my blood sugar? - ***≈40* mg/dl**

Time	Blood glucose [mg/dl]	+/- Urine glucose test strip	Action
Phase 1: Achieving the basal conditions (approx 2 - 3 hours)			
15 00	127	Spuren	Last meal at ______
15 30	115	neg.	Last injection at ______
15 50	107	neg	Last correction at ***≈12:40***
16 30	98	neg	With ***3U Humalog***
17 00	109	neg.	Because ***BG ≈ 240 mg/dl***
Phase 2 + 3: Raising blood glucose and reaching BG plateau (approx 2 - 3 hours)			
17 10			30g glucose 2 carb choices
17 20	141	neg	
17 30	148	neg	
17 40	179	neg	
17 50	174	neg	
18 00	198	neg	
18 10	224	! pos	
18 30	290	pos	
19 10	250	pos	
Phase 4: Controlled lowering of blood glucose (approx 2 hours)			
19 30	247	pos	3,5 U Humalog
19 45	210	pos	
20 00	184	pos	
20 20	170	Spuren	
20 45	121	neg	
21 00	98	neg	

Fig. 7.4b: Patient protocol (determining the kidney threshold of glucose)

Can the "kidney threshold" test be done by starting with very high blood glucose values, perhaps when the blood glucose is already above the kidney threshold?

Checking blood glucose corrective algorithms under these conditions is problematic. It would be better to consider repeating the kidney threshold game under more suitable circumstances. But, principally, in this case, you could carry out the test in reverse order, first intentionally lowering the blood glucose and then intentionally increasing it. However, with urinary glucose excretion evident basal conditions are not met, especially since you are continually losing glucose in the urine.

Are there any tips for checking the kidney threshold?

1. Do not carry out this exercise immediately after a day of fasting or after the so-called "Yielding to Temptation", since "basal" conditions are particularly difficult to reach then. The glucose production in your liver depends on available glucose "stock", from previous nutrition. The liver stock of glucose is relatively empty after fasting or very full after a "Celebration Day". An otherwise suitable basal dosage is then too low or too high for you.
2. Always use the same blood glucose meter during the kidney threshold experiment to avoid any problems with interpretation.
3. Use the most sensitive urine glucose strips possible.
4. Carry out the test in the presence of an experienced FIT counselor. You will need help to interpret the course of the test.

As already presented in the table at the end of the book, there are some simple equations to provide your initial correction algorithms. For increasing your blood glucose, if below the target, you can assume that 1 carb increases your blood glucose by _______ = 90 – (kg body weight x 0.5).

For lowering your blood glucose:
The algorithm for correction of hyperglycemia "1 unit of rapid-acting insulin lowers my blood glucose by ______ mg/dl" is dependent most on your total daily insulin need. An approximation to your algorithm can be derived with the following formula:

1700 : Total Daily Insulin Dose = __________ (round up) / mg/dl *

which can be used for people with daily total insulin consumption between 20 and 80 units and "acceptable" MBG of 150-280 mg/dl.

In an even worse glycemic control does this formula not apply?

If you are that high, insulin does not work as usual. As discussed for ketoacidosis, high values associated with dehydration and ketones lead to insulin resistance.

---

*Footnote: The equation "1500 : TDD" was provided by Dr Davidson in the US for regular insulin

Experience shows that with blood glucose values over 400 mg/dl you can even *double* your estimated insulin amount used for correction of hyperglycemia. However, the risk of over dosage must be taken then into account! Therefore, check your blood glucose more often and drink a lot of fluids in such situations. You can avoid the majority of such strong blood glucose elevations while correcting al minor increases immediately!

Are there any other factors which can influence blood glucose lowering while correcting hyperglycemia?

Rapid insulin analogs induce less blood glucose lowering than a longer acting, subcutaneously applied regular insulin. Even less effect will occur if using an exceptional intravenous access for insulin application while correcting high blood glucose values. Let´s summarize: shorter duration of insulin action is associated with less blood glucose lowering effect while correcting hyperglycemia. Practical consequence: Patients with history of hypoglycemia with unconsciousness should preferably use rapid acting analogs instead of regular insulin. Insulin analogs are statistically less associated with severe hypoglycemia. That could be due to reduced probability of overdosage while correcting on one hand, and maybe with better perceived and stronger symptoms of fast blood glucose lowering on the other one (when compared with slower and longer acting regular insulin).

7.3 Basal Insulin Replacement and the Fasting Day

Do I need to start fasting in order to determine my basal insulin requirement immediately?

Don't overestimate the value of the fasting day as a test to examine the adequacy of your basal dosage. Avoiding food for 36-hours between the evening meal and breakfast two mornings later reduces your glucose reserves (glycogen) in the liver. The liver directs glucose production so it sets the fasting insulin requirement by gauging the quantity of stored glycogen reserves. An otherwise suitable basal insulin dosage may be too high for a long-term fast of more than 20 hours.

Could I get hypoglycemia during long-term fasting even with a 'good' basal rate?

Basal insulin dosage is optimal when -- on the fasting day -- app. 1.5-3 carbs are necessary to stabilize blood glucose levels

On the day of the fast, you will not experience serious hypoglycemia as long as you control your blood glucose often. Measure your blood glucose approximately every 2-3 hours. Correct blood glucose values under 80 mg/dl with glucose to your target, as usual. In our experience, a basal insulin dosage is optimal when -- on the day of the fast -- approximately 1.5-3 carbohydrate choices maximum: 45g of glucose are necessary to stabilize blood glucose. Many people with diabetes have also discovered that on the fasting day, or if skipping breakfast and lunch, they need to reduce their "morning mound" rapid insulin by about one-half.

My "morning mound" basal rapid insulin consists of 4 units. On a fasting day would I inject only 2 units of rapid insulin in addition to my usual dose of delayed-acting insulin?

Yes, as long as your blood glucose is in the target level when you wake up, approximately 100-120 mg/dl. If it is higher, you have to inject rapid insulin to correct to your target, as usual.

What do I need to do if my blood glucose is 60 mg/dl when I wake up on a day that I plan to fast?

The blood glucose correction always has the highest priority. A value of 60 mg/dl is never a target value. You need to calculate with your FIT rule: "1 carbohydrate choice increases my blood glucose by ___ mg/dl". In other words, you need to know how much glucose to eat to increase your blood glucose to your target level. Then you can inject your usual amount of delayed-acting insulin while adding your "morning mound" of rapid insulin reduced to 50% of the usual dose. Alternatively, with low blood glucose values you can skip your "morning mound" rapid insulin instead of taking glucose (however: less safe!).

Does this spontaneously increase blood glucose?

The dawn phenomenon, as mentioned before, is caused by increased glucose production during early morning hours. Therefore, you need more rapid acting insulin in the morning or at breakfast. A fasting day provides you with an opportunity to observe your blood glucose values without breakfast. An increase in blood glucose is still expected to occur despite an injection of rapid insulin.

Now I understand why it is so difficult to achieve acceptable blood glucose values after breakfast if my blood glucose increases on its own.

Imagine that in the morning your liver delivers a particular amount of glucose into the blood stream in addition to your normal fasting glucose production. If you want to eat two rolls in the morning (four carb choices), you need to consider that your liver maybe produces one or two more additional "rolls" or another four "carbohydrate choices".

Does this mean that altogether around 8 carbohydrate choices will need to be transported into my cells?

Yes, it does. Technically, it is especially difficult when you already wake up with increased blood glucose values.

INTERNATIONAL RESEARCH GROUP
ON FUNCTIONAL INSULIN TREATMENT
Medical University Vienna
kinga.howorka@meduniwien.ac.at
www.diabetesFIT.org

Name: Emma M.
Birth date: 1949 Phone: ...
Address: ...
E-mail: ...
Diabetes since: 1964 Wt.: ...
FIT since: ... with ✓ injections O pump

I BASAL (fasting insulin): AM 12 LANTUS/3 APIDRA **N** PM 12 LANTUS U **S** PRANDIAL: for 1 carb choice (15g) = 1,5 APIDRA U **U**	**Target for correction of aberrant BG values:** Fasting/pre-meal: 100 mg/dl (or ...) After meals: 1h<180 (or...), 2h< 140 mg/dl **Target range for MBG: from ... to ... mg/dl**

L CORRECTION: 1 U rapid ins. lowers my BG by approx.- 40; 1 carb raises my BG by approx.+ 50 mg/dl

I EXAMPLE: carb choices/CHO: ...
N insulin (U): ...

	TIME	1	2	3	4	5	6	7	8	9	10	11	12	1	2	3	4	5	6	7	8	9	10	11	12	Total Daily Dose	
							AM													PM							
MON	Basal	Lantus						12															10			22	
	Bolus	Apidra						2			2															4	26
7	BG				92			127			154			107				64			81		64			MBG	98
Jun	Carbs/CHO																	0,5					1				1,5
	Comments	Fasting									corr.							1/2 carb					1 glucose				
TUE	Basal																										
	Bolus																										
8	BG				90			79																		MBG	
Jun	Carbs/CHO																										

Fig. 7.5a: Insulin game: "The Fasting Day". The average total daily insulin dose of Emma is usually between 50 and 55 insulin units (rapid- plus delayed-acting insulin/day). You can see the blood glucose course during fasting in her log book. What do you think of Emma's basal insulin dosage?

The Lantus® dosage needs to be:

- Increased (mornings and evenings)?
- Reduced (mornings and evenings)?
- Reduced only in the evening?
- Kept unchanged?

Fig. 7.5b: During these 36 hours fast, several units of rapid acting analog were used for blood glucose correction (in the morning). Later, however, glucose (in total 1.5 carbohydrate choices) had to be taken because of too low blood sugar values. Evening long acting insulin has been reduced by two units for safety reasons.

This experiment of fasting 36 hours is much longer than usual fasting periods between meals (the longest being of 16-18 hours over night). In such a case the sugar depots are consumed and the glucose production of the liver reduced. Patients' experience shows that the correct basal rate will require up to 2-2,5 carbohydrate choices to maintain blood glucose stability on a fasting day and "keep up with the basal". With fasting blood glucose usually between 90-130 mg/dl and an average awareness of hypoglycemia, the basal insulin dosage can therefore be left here unchanged.

An even lower basal rate is nevertheless to be considered when:

- planned weight loss
- repeated serious hypoglycemia (including loss of consciousness) in the past

I have two remedies. I can prolong the interval between injecting and eating, if regular insulin and not rapid acting analog is used, or I can accelerate the absorption of regular insulin, for example by injecting into the muscle. I am happy that I do not have to eat breakfast on a fasting day since I am usually not hungry in the morning. Until now, I always had to force breakfast down. Can I really go without eating breakfast in the future?

Nothing is stopping you from skipping your breakfast! Eat when you want, what you like, and as much (or as little☺) as you wish and compensate accordingly with insulin.

Returning to the fasting day, does the 36-hour fast affect the future basal insulin dosage?

The basal insulin dosage is correct when only a few carbohydrates (maximum of 2-3 carbohydrate choices) are consumed on a fasting day to avoid hypoglycemia. These carbohydrates are eaten "to keep up with the basal rate", without requiring rapid insulin. If blood glucose increases on a fasting day apart from the morning blood glucose correction, and other corrections with rapid insulin are necessary, then the basal insulin dose must be increased. Conversely, when large quantities of carbohydrates are consumed because of blood glucose decreases (more than 45 grams or 3 carbohydrate choices) to stabilize blood glucose, then the basal dosage needs to be reduced. *Do not* go to sleep with very low blood glucose values. Increase the values late before going to sleep to over 120 mg/dl. People who tend toward serious hypoglycemia need to increase the values to over 140 mg/dl.

Could you please summarize, how to replace best basal insulin requirement?

Basal insulin transports glucose produced from the liver while fasting to the cells. The basal insulin need is not homogenous throughout the whole day! It increases in the morning ("dawn" phenomenon) and in the late afternoon ("dusk" phenomenon); at lunch time and several hours after midnight your basal insulin need is usually low. These circadian fluctuations can be best compensated with an insulin pump, delivering different amounts of short acting insulin accordingly.

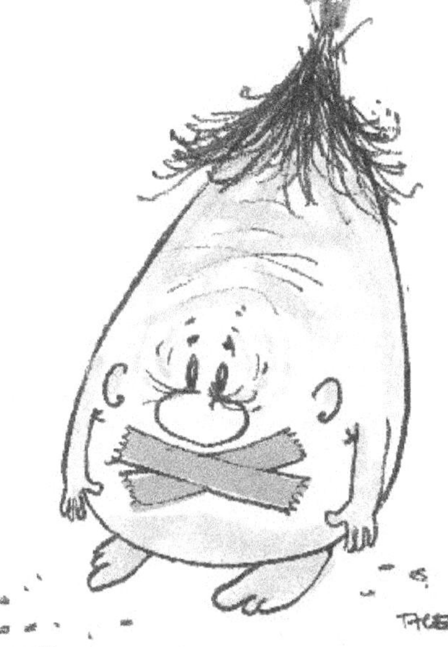

What if I don't want to have a pump?

Don't worry. So far, you have learned the profile of delayed-acting insulin available: The basis can be set with NPH insulin (rather 3 then 2 injections per day), as well as insulin Levemir® (2 injections per day) or with insulin Lantus® (1-2 injections per day). If high fasting values indicate pronounced dawn phenomenon, a "combined" basal with Lantus® in the morning and NPH- or Levemir® in the evening is usually effective for better fasting values, comparable to those reached otherwise only with pump treatment.

It is my understanding that a bit of Lantus® in the morning allows stability throughout the day, while a bit of Levemir® late in the evening hinders the "dawn" phenomenon? Both insulin preparations do work in the early morning hours then, don't they?

That is correct. Such a "combined basal" is therefore quite effective for controlling high fasting values.

Dawn phenomenon is induced by an increase of growth hormone while falling asleep. It can be increased as well:
- In the puberty,
- With weight gain,
- After an excessive supper/dinner, increasing storage/reservoirs of the liver.

Dawn phenomenon diminishes by:
- Glycogen storage depletion in the liver (fasting, "dinner cancelling") ,
- Longer diabetes duration,
- The end of puberty,
- Exercise with consumption of glycogen storage,
- At high dosage of delayed-acting insulin ("suppression of liver") for basal purposes.

Circadian fluctuations in basal insulin need tend to decrease with longer diabetes duration, probably as a sign of an incipient autonomic neuropathy.

Dusk phenomenon, an increase in insulin need late in the afternoon, is induced by an increase of cortisol in the early morning mirrored late afternoon/evening with the increased production of glucose. It is increased:
- In pregnancy,
- In the 2nd half of menstrual cycle (progesterone phase),
- Overweight,
- Cortisol/ prednisolone therapy.
- Too low basal rate.

On a fasting day you'll experience a reduction of the "dusk" and "dawn" phenomenon. When the liver becomes empty, circadian fluctuations decrease. Consequently, the basal dosage can be reduced by 10%. Alternatively, you will need 2-3 carb choices usually to be taken after approx. 20 hours of fasting.

Let´s come back from fasting to our daily routine. For the assessment of the correctness of your basal choice, you have to take three criteria under consideration:

1. Stability of blood glucose after short-term fasting between meals and over night;
2. Fasting values (preferably 90-130mg/dl);
3. Proportion of delayed-acting insulin (basal rate) in total daily dose. Delayed acting insulins should only seldom exceed 50% of total daily dose.

Is it really worth to fast in order to test the basis?

Not for the sake of basal dosage, since the criteria mentioned above are more important. Furthermore: the amount basal insulin you need at the moment (=how much glucose is produced in your liver and intestines at the moment) greatly depends on the amount of food recently eaten.

So, why should we fast at all?

First: because you'll surely have to fast one day or the other: e.g., due to a surgery, anaesthesia, birth, gastroscopy, vomiting and diarrhoea or weight reduction. It would be unreasonable, not to learn it now. At this moment, use the opportunity of the supervisor during the FIT program.

Second: this "game" helps to raise your awareness and make you always realise – in the case you have a type 1 diabetes – that under no circumstance should you completely skip the basal insulin, even if you haven't eaten anything e.g. due to vomiting.

By the way, it is quite healthy to fast. We have shown and published that fasting considerably improves the function of autonomic nervous system (that is responsible, among others, for erection ability and cardiovascular health☺).

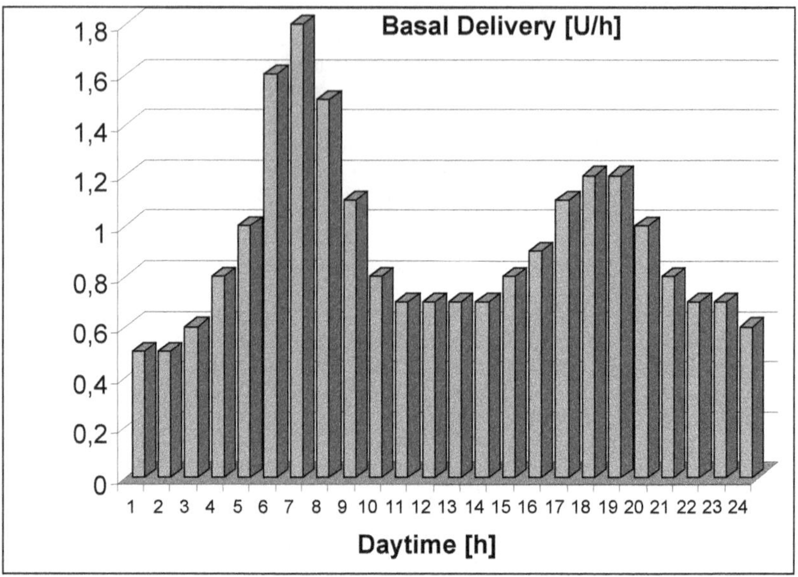

Fig 7.6. Typical fluctuations in circadian basal insulin requirement (here, the total sum equals to 22 U/day) in adolescents (determined empirically during CSII/insulin pump therapy; distinctive dawn and dusk phenomena; modified after Danne et al., Pediatr Diab, 2006: 7, Suppl 4, 25-31).

"Dawn phenomenon" is the term describing an early morning blood glucose increase, due to growth hormone, which is produced while falling asleep but raises blood glucose only several hours later, at dawn. An additional morning rise is due to the immediate action of adrenalin while standing up ("standing up" phenomenon). Cortisol on the other hand, is produced in the morning, but like growth hormone, it takes several hours to increase glucose production only in the late afternoon. Cortisol is thus responsible for the so called "dusk phenomenon", late afternoon blood glucose increase. Circadian fluctuations disappear with longer diabetes duration. Dusk phenomenon is elevated during pregnancy, whereas in puberty the elevation of dawn phenomenon is typical.

8. Hypoglycemia: Low Blood Glucose

Is it true that brain cells die during hypoglycemia?

Not with minor hypoglycemia. Animal experiments show that sustained brain damage can only occur with very serious hypoglycemia, with convulsions. In treating people, we all know that serious hypoglycemia with loss of consciousness needs to be avoided. Major hypoglycemia may affect hypoglycemia perception and decrease awareness of low blood sugars. Thus, it needs to be avoided if possible.

What about minor hypoglycemia, is it harmless?

Not quite: if it occurs more often, it could attenuate hypo symptoms, meaning it lessens one's awareness of hypoglycemia.

What are the symptoms of hypoglycemia?

Blood glucose values below 60-70 mg/dl can lead to a temporary relative lack of glucose supplying the brain. Intellectual abilities weaken when the blood glucose fall below 60-70 mg/dl (even if the symptoms of hypoglycemia are not obvious), since glucose is the exclusive fuel for the brain cells. Hypoglycemia can lead to

- a lack of concentration,
- nervousness,
- irritability,
- weakness,
- perspiration,
- fatigue,
- headaches,
- shakiness,
- and dizziness.

People usually experience some, not necessarily all, of these symptoms but some people do not have any symptoms until they are experiencing serious hypoglycemia. To compensate for this lack of fuel, the body tries to increase blood glucose using various mechanisms. The blood glucose is increased even without carbohydrate intake by counter-regulatory hormones releasing glucose from the available glycogen stores in the liver. These immediately acting hormones include adrenaline and glucagon. The delayed acting hormones are cortisol and growth hormone. These prolonged acting hormones are also released during stress. All the classic symptoms of hypoglycemia such as sweating, shaking and heart palpitations are caused by the discharge of adrenaline, the most important stress hormone. The brain's functioning is affected leading to confusion and finally lack of consciousness when the body fails to reverse hypoglycemia and blood glucose continues to fall.

What is the difference between "minor" and "serious" hypoglycemia?

The criteria of hypoglycemia are determined by how much of the brain's functions are disturbed by the lack of glucose. The absolute hypoglycemia level of blood glucose does not always mirror perceived symptoms. According to the functional condition of the brain, the following degrees of hypoglycemia can be distinguished:

1. **Minor hypoglycemia (H1):** symptoms of hypoglycemia with blood glucose values under 70 mg/dl or accidental blood glucose values free of symptoms under 60 mg/dl. No noticeable limitation of intellectual abilities exists with mild hypoglycemia, apart from how unpleasant or dramatic a "hypo" might be (you might perspire heavily or not). As long as you can still think clearly, hypoglycemia is rated as "minor".
2. **Medium-serious hypoglycemia (H2):** characterized by loss of rational reasoning. This occurs when you are physically present but mentally "absent", i.e. you are "spacey". It is essential to avoid this dangerous condition of "mental confusion" by taking glucose with early symptoms of H1.
3. **Serious hypoglycemia (H3):** involves loss of consciousness. In this condition, someone with diabetes is completely dependent on others for help.
4. **Serious hypoglycemia requiring medical intervention (H4):** loss of consciousness treated with glucagon or with intravenous glucose.

It is important that glucagon is kept at home, school or workplace of every insulin-treated person

What is glucagon?

Similar to insulin, glucagon is also a protein hormone produced by the pancreas, but its effects are completely opposite to insulin: glucagon increases blood glucose. It releases glycogen i.e. glucose depots in the liver. It must be injected like insulin because glucagon is a protein. Glucagon can be injected under the skin, or into the muscles, intramuscular, or intravenous. Your family members need to be able to inject glucagon in the event that you ever become unconscious. They need to know that loss of consciousness in a person with diabetes is highly likely to be caused by hypoglycemia. Injecting glucagon is easy and safe. Keep in mind that your family members have less experience with injections than you. You need to tell them how glucagon is given and practice it with them (Fig. 8.1)

My relatives are always afraid that if I become unconscious, they would not know whether it was due to low or high blood glucose.

Both coma states related to hypoglycemia and ketoacidosis, are easily distinguished from one another. Serious hypoglycemia is much more likely from a statistical point of view. Apart from the statistical probability, it is important to know that serious hypoglycemia can seem to occur "out of the blue", but ketoacidosis is usually associated with a serious illness and needs time to develop.

Is it true that people on FIT rarely lose consciousness because of overly high blood glucose?

That's right. If you comprehensively carry out blood glucose self-monitoring several times every day and immediately correct each, even a slight, blood glucose increase, ketoacidosis is unlikely, without an acute illness being present, of course. For family members who are less informed, deliver the pacifying message that it would be difficult to slip "high" because you carry out self-checks and corrections so frequently.

Other symptoms are recognizable with serious hypoglycemia with loss of consciousness, like a drop in body temperature, cool skin, and increased perspiration related to stress hormones like adrenaline. These symptoms clearly differentiate this condition from a diabetic coma. A diabetic coma is characterized by signs of dehydration and "forced" breathing ("Kussmaul's breathing" is a compensatory deep labored breathing pattern for severe metabolic acidosis) induced by an increased acetone build-up in the blood, the existence of acids - ketoacidosis.

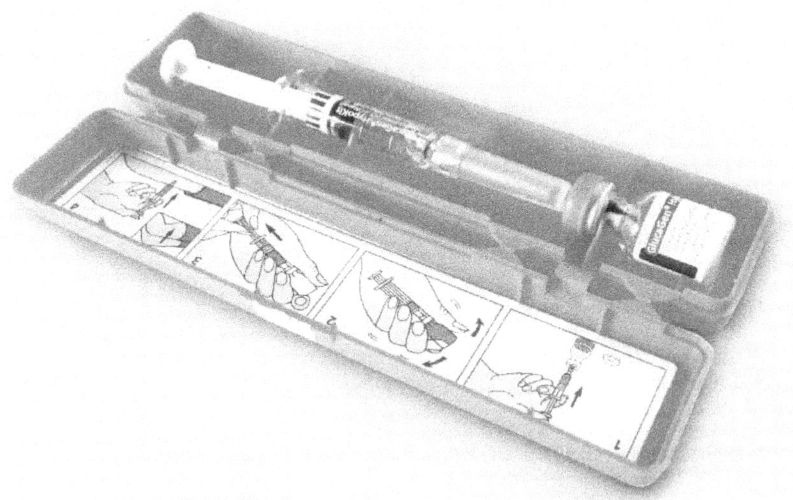

Fig. 8.1.: Glucagon is a white powder that needs to be diluted in the diluent before injecting.

After injecting the solution into the bottle it mixes immediately with the powder together. The whole content (1 mg) of the glucagon vial (the volume of the diluting liquid is not so important) will be injected. After becoming conscious again, it is important to eat some carbs in order to avoid another fall in blood glucose.

Minor hypoglycemia will happen more often than ketoacidosis. Should I immediately eat carbohydrate rich foods?

Yes. If you are unsure whether it is really hypoglycemia then test your blood sugar. You will overshoot the target if you eat too much to correct hypoglycemia. Use only rapidly absorbed carbohydrates or simple sugars. Glucose tablets are the best. One tablet contains approximately 4 grams of glucose. Sugar is taken more quickly from the stomach and intestines when you drink liquids because increasing the volume of mass in the stomach accelerates absorption. Also, soft drinks containing sugar or fruit juices are very good for treating minor hypoglycemia.

Do I need to use slowly absorbed carbohydrates when treating minor hypoglycemia?

Not with FIT. Hypoglycemia either exists or it does not. If very low blood glucose values or hypoglycemic symptoms exist consider the motto "treat immediately or not at all." A comparison exists for correcting high blood glucose with insulin. Or, do you want the "counter-regulation" of your body to be activated?

Formerly, when experiencing hypoglycemia, I was always given chocolate for hypos.

Chocolate is not the best for treating hypoglycemia because the fat content delays the absorption of carbohydrates. It takes too long. The eventual blood glucose increase (depending on the amount of chocolate consumed) maybe will be very high, after a couple of hours, but the effect takes too long. If you want to increase your blood glucose immediately avoid fatty foods and eat simple sugars.

Do I need to carry glucose tablets with me at all times?

Please do, since anyone who injects insulin can get hypoglycemia. If you want to achieve good glycemic control, never underestimate the risk of hypoglycemia.

I eat immense amounts of food when I have hypoglycemia, I just can't control myself! I feel so hungry!

One possible reason for your great hunger is that you may have eaten foods containing complex carbohydrates which are absorbed too slowly. Consequently, it can take a long time until your blood glucose increases to a point where you feel in control again. Or, you may not have been prepared in advance for hypoglycemia and did not have glucose tablets with you. Should this happen again, eating too much to correct hypoglycemia is not the worst scenario. Do not feel guilty. If you consume seven carbohydrate choices when your correction needs are only two carbs, compensate for the other five carbs with insulin. In this case inject insulin after your blood glucose is normal again to avoid further increase. Be prepared next time and remember to always carry along glucose tablets with you

I doubt that anyone could always avoid hypoglycemia.

You are right. However, several options are still at your disposal to lessen the frequency of hypoglycemia. This is relevant to you now and in the future, since you have made the decision to maintain blood glucose control yourself.

I know people with diabetes who rigorously perform self-monitoring and seek treatment but still have serious hypoglycemia frequently.

Unfortunately, certain people are prone to serious hypoglycemia. They need to use particular preventive measures that will be discussed later. You can recognize whether you have an increased risk for serious hypoglycemia.

Hypoglycemia risk patients:

- have experienced hypoglycemia with loss of consciousness in the past, or were confused at times,
- display often blood glucose values below 50 mg/dl indicating some degree of hypoglycemia unawareness,
- have had diabetes for a long time,
- have impaired kidney function,
- are very thin,
- are very ambitious with their diabetes management, which results in HbA_{1c} values being quite low.

The probability of serious hypoglycemia in the future is particularly high if you have several of these characteristics. According to our experience, approximately 10-20% of all people with type 1 diabetes have such increased risk for serious hypoglycemia.

What particular safety measures do these people need to take?

Most importantly, they need to be aware that they are *at high-risk*. A specific hypoglycemia prevention group module is often necessary. They need different alternatives to implement therapy. They also need to know what situations prompt hypoglycemia.

Patients with high risk of hypoglycemia need special preventative measures

What are these situations?

Hypoglycemia risk situations are induced by not recognizing the symptoms of hypoglycemia. Following are typical examples:

- Exhaustion.
- Alcohol consumption because it inhibits the production of glucose in the liver.
- Not eating enough food. The glucose reserve in the liver depletes. Weight control diets or fasting require a dramatic reduction in the basal rate.
- Certain medication such as beta-blockers (high blood pressure treatment) because they weaken your body's response to hypoglycemia. Sympathomethics (asthma treatment) because they may imitate symptoms of hypoglycemia causing confusion about whether low blood glucose actually exists.

- Skipping or omitting glucose tests. "Flying blind" is dangerous. It is important not to correct high glucose because you "feel" high. Test you blood glucose before correcting it!
- Eating food with an unknown amount of carbohydrates, especially sweets because there is the danger of misjudging prandial insulin requirements and thus, over-dosing insulin.
- Exercising without food and/or insulin dosage adjustments.
- No adequate diabetes education. This is dangerous because of "unprofessional" use of insulin.

What factors contribute to hypoglycemia?

On FIT, there are only three reasons for hypoglycemia. They appear separately or in combination:

1. Too much insulin for *fasting;*
An excessively high basal rate will occur temporarily when your carbohydrate intake is low, e.g., during physical exercise or alcohol consumption. Too much basal insulin can, at times, be available when delayed-action insulin has been wrongly distributed over the day (ie., once daily Lantus® injection for basal needs). If you take only *one* injection of basal insulin such as Lantus®, you have "unevenly" covered your basal need. Very important rule to be reminded: never use more than half of your total daily insulin requirement as delayed acting insulin.

2. Too much insulin for *food eaten*;
   a) Overestimating the amount of carbohydrates in a meal is the most common reason for hypoglycemia. In the "training" phase for FIT it is important to note the insulin dosage injected for each meal/snack to be able to estimate the correct insulin to carbohydrate dosage for the future.
   b) Unpredictable insulin absorption due to lipohypertrophy of the skin (i.e. skin areas with fat overgrowth due to multiple injections in the same spot) caused by injecting insulin in the same area again and again.
   c) Mixing delayed acting insulin with rapid insulin can delay the insulin effect.
   d) Delayed stomach emptying due to gastro-intestinal nerve damage, sometimes observed in patients with late diabetes complications. Special precautions need to be taken to avoid hypoglycemia.

3. Too much insulin for a blood glucose *correction.*
Blood glucose corrections with rapid insulin can lead to hypoglycemia when the wrong target point is selected. In hypoglycemia risk situations, a higher target for corrections needs to be temporarily selected. People at risk of hypoglycemia need to have a higher target, a pre-meal target of approximately 120 mg/dl or over and a post-meal target of 200 mg/dl. Post-meal glucose corrections, approximately 2 hrs after prandial insulin, are relatively risky since the insulin to carbohydrate absorption speed is not always predictable. Post-meal, a blood glucose correction can only be carried out to a higher post meal target point, and only if no correction was made with the prandial injection. Avoid "double corrections." To avoid hypoglycemia, it is advised that the "shortest" interval between two consecutive high blood glucose corrections no

less than 3 hours if using regular insulin and no less than 2 hours if using rapid acting analog.

Are there any other ways to avoid hypoglycemia?

Yes. Always use as little insulin as possible. Long-acting insulin should never cause hypoglycemia. You need to establish a basal rate so low that it promises just acceptable blood glucose values when fasting. Basal should only seldom exceed 50% of total daily dose. Inject rapid-acting insulin separately for each particular meal or snack containing carbs.

I heard that there are new sensors that warn patients with diabetes of hypoglycemia.

As long as continuous glucose measurement is not everybody, this would be a major leap forward. Hypoglycemia risks present the greatest limitations to effective insulin therapy. At the moment, there are companies trying to develop practical equipment for people who cannot detect hypoglycemia sufficiently, an alarm system so to speak. However, until such equipment is available you need to try to develop your own "sensor" to detect hypoglycemia.

Me, by myself?

Sensitive, practical tests for detecting *impaired rational thinking* are meaningful. The nervous system, particularly the brain, reacts sensitively to blood glucose decreases. Glucose represents almost the exclusive energy source for the brain. Your ability to think rationally decreases when blood glucose levels are 50-65 mg/dl or lower. This is long before the classic symptoms of hypoglycemia occur such as sweating, shaking etc. You can prove it. If you were aware of your impaired condition you could treat it. People have reported to us that hypoglycemia awareness is improved with training. Many people with previous severe hypoglycemia show a typical symptom or "precursor" long before other symptoms appear. Particular emotions like fearfulness or depression that is not related to current situations are likely symptoms. A common symptom of hypoglycemia is confusion, so a memory test can be a good indicator or mild hypoglycemia. Here's how: What is 7 x 14? Can you recite "Twinkle, Twinkle Little Star" or your own telephone number? If you cannot or if they take too long, this may be a warning sign of

hypoglycemia. You need to test your blood glucose immediately and take appropriate steps to bring it to your target values.

You said that perfectionists are at risk of developing hypoglycemia. Why?

This trait characterizes people who seek constant "perfect" values instead of a more realistic fitting blood glucose "target area" to life circumstances. The current blood glucose target should be chosen according to special situations to maintain the best glycemic control with the lowest risk of serious hypoglycemia and minimal effort. Having MBG values below the range of 90 mg/dl is preferable to be achieved by pregnant women in the last trimester, otherwise for all the other patients another goal should be set: not pregnant "hypo endangered" patients should strive to achieve MBG values over 160 mg/dl, where Hba_{1c} values should not be less than 7,5%.

It has been demonstrated that repeated episodes of mild hypoglycemia, values of 50-70 mg/dl, lessen a person's awareness of the symptoms of hypoglycemia. This is specifically related to people who preserve to stay "low". For these people, it becomes more difficult with time to recognize symptoms of hypoglycemia.

In contrast, a different study showed that hypoglycemia awareness may be reestablished by systematically avoiding mild hypoglycemia. This is a big consolation to people with high hypoglycemia risk. They can convince themselves to change their behavior. For these people FIT has two special training sessions to address specific re-establishment of hypoglycemia awareness. Our experience demonstrates that such "hypoglycemia group module" should be standard training for people at risk of severe hypoglycemia.

My hypoglycemia awareness seems to vary from day to day....

This is correct, but it seems that altogether it is usually quite individually specific. As already mentioned, about 10-20% of type 1 diabetic patients, those with history of repeated unconsciousness, have considerable hypoglycemia unawareness. On the other hand, however, another 10-20% of patients display supranormal hypoglycemia perception. Typically, these patients have never been unconscious even after long diabetes duration.

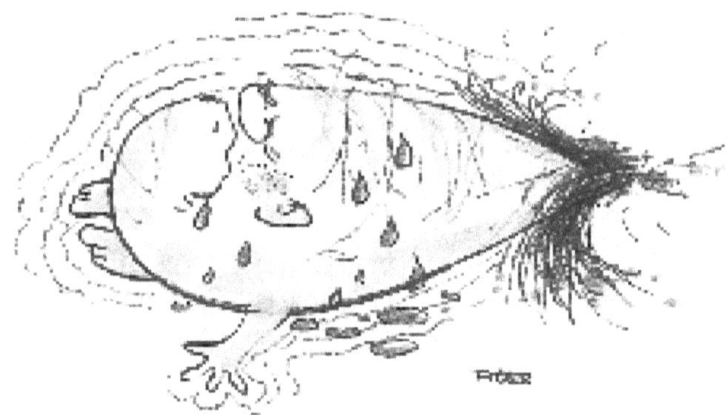

"Supranormal" hypoglycemia perception?! It must be awesome to perceive each slightest sign of a hypo!

No, it's not as great as you think; if you associate each hypo with very severe and very disturbing symptoms, you will tend to keep high all the time to prevent such unbearable states. That makes a sensible treatment unachievable..., sometimes with the consequence of much earlier and more severe (micro vascular) complications. Experience shows that such 'supranormal' hypoglycemia perception with severely disturbing symptoms can sometimes be attenuated by a trial of pump treatment.

Any tips for those endangered with severe hypoglycemia?

Some people hate their glucose tablets. (Semi-) fluid products specifically developed for insulin reactions sometimes taste awful. Try sweet condensed milk – it tastes in hypoglycemia apparently "without resistance".

Any treatment hints for hypoglycemia unawareness?

To summarize: The most important is avoidance of even slight hypoglycemia. Usually it will result in higher average glucose as discussed already.

Are there any more tips for perfectionists?

Perfectionism in diabetes control is often associated with correcting too frequently or too dramatically. If you exhibit such perfectionist tendencies you can intentionally increase the algorithm: "1 unit of rapid-acting insulin lowers my blood glucose by ____mg/dl" to avoid an insulin overdose. For example, instead of "1 unit of rapid insulin lowers my blood glucose by -40 mg/dl" increase this to "1 unit of rapid insulin lowers my blood glucose by -50mg/dl." The quantity of rapid insulin to be injected for the correction of blood glucose is now "less."

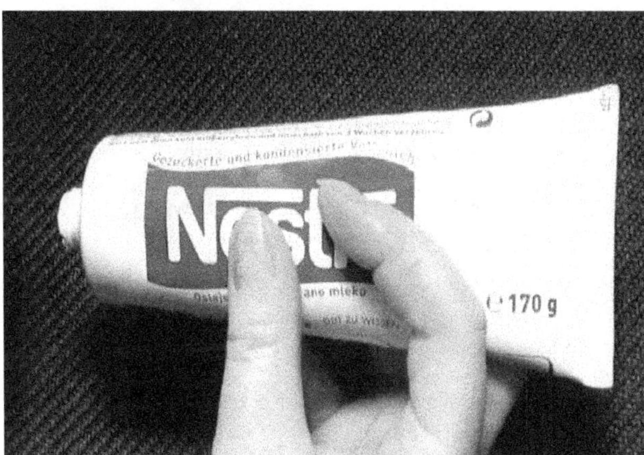

Fig 8.2. Sweetened condensed milk
Tastes in hypoglycemia apparently "without resistance".

Please summarize hypoglycemia prevention steps.

Do not attempt FIT without intensive training. Functional therapy only stays "near" normal glycemia for safety reasons. Striving for real normoglycemia, blood glucose values between 60 and 120mg/dl, is not advantageous or safely possible using today's methods. Striving for such real normoglycemia is acceptable only for pregnancy. If you are not pregnant, HbA_{1c} values within 1-2% above the upper reference range is probably safe enough to avoid late complications in the majority of cases. If the lab's HbA_{1c} upper reference range for normal is 6.1 an HbA_{1c} of 6.7-7.3 would be "ideal." A higher target level for hyperglycemia correction must be chosen for people who are prone to serious hypoglycemia. Mean daily blood glucose below 130 cannot be recommended to people at risk for serious hypoglycemia, MBG of 130-200 will usually be more appropriate if you had a history of unconsciousness due to hypoglycemia.

The following steps can be taken to avoid hypoglycemia with FIT:

1. Prepare in advance for hypoglycemia. Learn how to recognize "hypoglycemia risk situations" in everyday life. Always test your blood glucose before driving a car. Always carry quick carbs like juice or glucose tablets with your "minimal equipment".
2. Always carry your meter and your FIT-minimum equipment with you to test and correct your blood glucose at any time.
3. Take as little long-acting insulin as possible. The basal insulin is not supposed to cause spontaneous decreases in blood glucose during short-term fasting. Fasting values under 90 mg/dl generally indicate too much basal insulin, more than 50% of basal in total daily dose indicates usually too high basal dose. People with history of severe hypoglycemia with unconsciousness should not use the "combined" basal rate including NPH insulin late in the evening as it could enhance hypoglycemia.
4. Take as little rapid acting insulin as possible. The best preventive measures are to inject at each meal and only to correct your blood glucose at minimum 3-hourly intervals. It is extremely important to avoid double corrections in shorter time periods.
5. Select a safe blood glucose target. In certain people with diabetes, and in specific circumstances, higher blood glucose correction target points must be chosen.
6. Take rapid acting insulin analogues instead of regular insulin. They reduce the probability of serious hypoglycemia by half. This is essential for those with high hypoglycemia risk.
7. Correct low blood sugar values "upwards" to your target with rapidly absorbed carbohydrates, even when no symptoms of hypoglycemia are felt.
8. If you tend toward perfectionism, establish an ambitious aim, "no serious hypoglycemia reactions" instead of "the best possible control." Read this chapter again in the next few days. Avoid even slightest hypoglycemia to improve hypoglycemia awareness – for the next few days, never ever leave a glucose level below 90 (!) uncorrected.
9. Learn the warning signs of hypoglycemia. A fast way to test yourself is to try to recall a poem, prayer, your telephone number or do a simple math task since an early symptom is confusion.

10. Keep fluid sugar, or sweetened condensed milk and glucagon at home, at work or school. Educate your family members, friends and co-workers on how to use it in case you become unconscious.
11. Discuss with your FIT professional particular preventive measures for your individual case. For disturbing hypoglycemia symptoms and hypoglycemia perception you may try a pump.
12. Learn from every hypoglycemic reaction. Stay calm and reflect. Why did it happen? What consequences does it have for the future? Is immediate raising your blood glucose enough or do you have to change your insulin algorithms?
13. If you experience serious hypoglycemia in the future review this chapter again! Seek help from your experienced FIT professional or physician.

Once again: Striving for real normoglycemia with blood glucose values MBG<140, to be set as a goal, is not suited in patients who have suffered loss of consciousness more than once. Goal for HbA1c: please do not strive for the values below the range of 7,5%. Even higher values of HbA_{1C} may be obtained if blood glucose target increase could be compensated with low blood pressure. To improve your prognose it may require a specific hypertension training and treatment.

9. Hyperglycemia: High Blood Glucose. Insulin Deficiency, DKA – Diabetic Ketoacidosis

Hyperglycemia means high blood glucose, right?

That's correct. It is the opposite of hypoglycemia. Hyperglycemia occurs when your blood glucose exceeds your selected blood glucose target level. If you determine that your current blood glucose is more than 40-50 mg/dl over your pre or post meal target level, you need to perform a blood glucose correction, i.e. use a correction bolus when more than one unit of rapid-acting insulin is necessary.

On FIT, I can rapidly treat high blood glucose, or hyperglycemia, at any time by immediately injecting rapid insulin. Correct?

Yes, although there is the double correction exception. The last correction of a high blood glucose value with rapid acting insulin needs to have been about 3-4 hours ago.

With blood glucose above target, even if already corrected, I need to wait a little before my next meal until my blood glucose is normal again. I think the recommendation is that I should not eat when my blood glucose is increased, I need to lower it first.

That's correct.

Do not eat if your blood glucose is too high; correct it first with correction bolus

If only I knew this before! I would have been able to prevent emergency visits to the hospital. I used to be hospitalized often because of very high blood sugars when I got sick.

An acute correction bolus may not be enough for "sick days": fever, flu, surgery. These can raise your whole daily insulin requirement and there are specific recommendations for "sick days"

So, I need to balance my insulin consumption every day, and when I see blood glucose corrections throughout the day, I will know I need more insulin than usual even though I eat less.

You need to watch for a continued increase in insulin requirements for a while with a bad case of the flu or other illness.

If I generally need about 50 units of insulin per day of delayed and rapid-acting insulin together, what will my insulin requirement be when I have the flu and a fever?

It is often about twice the usual daily insulin requirement for most people.

So instead of 50 units I shall need about 100 units a day?

True, but of course no one can predict this applies in all cases. You might take more rapid insulin for corrections and then note how the total daily dose of insulin has changed. In this way, you can easily confirm, that your daily need is e.g. doubled.

Let's say that I injected 24 units of delayed-acting insulin each day; 12 units in the morning and 12 units in the evening. But if my total daily dose doubles I need to increase the dosage for the basal rate immediately as well?

Yes you do, otherwise you would be constantly correcting your blood glucose. The basal rate, which was acceptable until now, will not be enough when your overall daily insulin requirement increases because of illness. You will also need to increase your insulin dosage for carbohydrates.

So, when my insulin requirement increases because of the flu, fever, or a similar situation I need to increase my insulin dosage for fasting and eating.

That's right. The question is to what degree to change your algorithms.

This depends on how much my insulin requirement has increased in total. In my particular case it doubled. Previously, I needed about 50 units and now I need an average of 100 units of insulin each day. Should I double the algorithms for fasting as well as for each carbohydrate choice?

Yes, you should.

Now I understand why it is so important to total insulin dosages every day. I must know how much insulin I need on an average day to adjust my insulin algorithms safely...

Correct! Only by totaling your daily insulin dose will you notice when your insulin requirement changes dramatically. For example, when you have flu, you need to expect variations in insulin requirement of approximately 20% even with a consistent lifestyle.

An experiment in type 1 diabetes examined insulin requirements on consecutive days. The insulin requirement in many of the patients varied greatly from day-to-day despite identical conditions, routines and diets.

Does this mean that I cannot prevent blips in blood glucose, even when I do everything right?

Exactly. You may not be able to prevent blips but now you can correct them safely! Do not overestimate the importance of these occasional blips. Frequently, in type 1 diabetes, several daily blood glucose corrections are necessary, even if you do everything "right". You do not need to concern yourself about it - it does not matter.

I always have a guilty conscience when I have high blood glucose values.

A guilty conscience does not help. You need to correct it and take charge of the situation. Keep these measures in mind:

1. Estimate your mean blood glucose (MBG) each day to evaluate your glycemic condition.
2. If you calculate a daily MBG over 160-170 mg/dl on several consecutive days, try to check if your insulin dosing algorithms are off.
3. Clarify whether your insulin requirement has changed for whatever reason and whether your insulin dosage algorithms need to be adjusted.

My activity levels change each day and I do not eat the same food everyday, even under similar conditions. So, I can assume that my blood glucose will probably fluctuate even more...

That's right. Try to differentiate between the following causes for increased blood glucose (hyperglycemia):

1. The administration of bolus forgotten (prandial or correctional).
2. Underestimating the absorption speed of carbohydrates from the bowel into the blood causes temporary hyperglycemia shortly after eating. This occurs when the inject-eat interval is too short and you have not accelerated the absorption of Regular before eating carbohydrate-rich foods. Alternative: rapid analogs.
3. Underestimating the amount of carbohydrates in a meal. In other words, the dose of rapid insulin for the meal is too low. You can only identify this several hours after eating, usually 4-5 hours later.
4. Morning hyperglycemia, due to insufficient basal dose or the so-called "dawn phenomenon."
5. Wrong algorithms for prandial insulin dosages.
6. Infections, flu, surgery, 2nd half of menstrual cycle, and other situations which cause an overall increase in insulin requirement.
7. Hyperglycemia following hypoglycemia, "rebound" or “Somogyi Phenomenon”. These rebound increases in blood glucose due to hormonal "counter-regulation" occur rarely with FIT and correct use of delayed acting insulin.
8. Technical malfunction, pen operating error, breaking a vial, clogging of a needle. Patients using pumps can be confronted with additional specific, mostly technological problems, such as:
   - empty battery, failure or breakdown, alarm ignored or too silent to hear
   - reservoir empty / glass ruptured
   - catheter not properly put together or not properly inserted; air in the catheter or clogged up

- canula disjointed, without noticing /return flow of insulin along the needle inserted into the skin

I see, in other words, the more complex the technology becomes the more prone it is to malfunctions...

True. Some patients have been using simple BD MicroFine- insulin syringes while eliminating technological errors to the bare minimum.

I realize that, usually, I will not need to change insulin algorithms after I consider these causes of hyperglycemia. Correcting my blood glucose will generally be enough?

That's correct. Without much consideration you can always bring your blood glucose into the chosen target level with an accurate blood glucose correction with rapid insulin. After thorough review, you need only to carry out a secondary change of your insulin algorithms when your average insulin requirement is higher over several days in a row. It will be evident that your basal rate is too low. We will return to this subject in the chapter about changing algorithms.

I do not quite understand how come so many diabetic patients end up in a hospital as a result of "severe highs" or ketoacidosis?

So called ketoacidosis occurs quite often as a result of underestimating the amount of insulin needed. Doubling total daily dose is sometimes necessary during infections. Moreover, a lot of patients are not used to checking for ketones in urine while having increased blood glucose values.

Why is testing for ketones so important?

Corrections of severe high: if your BG is over 400 mg/dl use double amount of short insulin for correction and check afterwards

If the blood glucose is increased and ketones excreted then this could only mean that there is a total lack of insulin. Remember: this is called the *"DKA": diabetic ketoacidosis.* Especially dangerous in the pump therapy, since there is an immediate and severe lack of insulin, should there be a disruption in supply. Not only severe highs but much more the ketones contribute to the development of DKA through acidification of the body. In order to prevent DKA effectively, you should always have at home ketone strips for urine testing. Don't forget, all severe highs can happen without warning. Insulin has a much weaker effect while acetone is being produced: thus double your estimated amount of short acting correction insulin, if your blood glucose exceeds 400 mg/dl.

Easily said: If your blood glucose is over 400mg/dl, use double amount of short insulin to lower it and check the BG afterwards, better repeatedly.

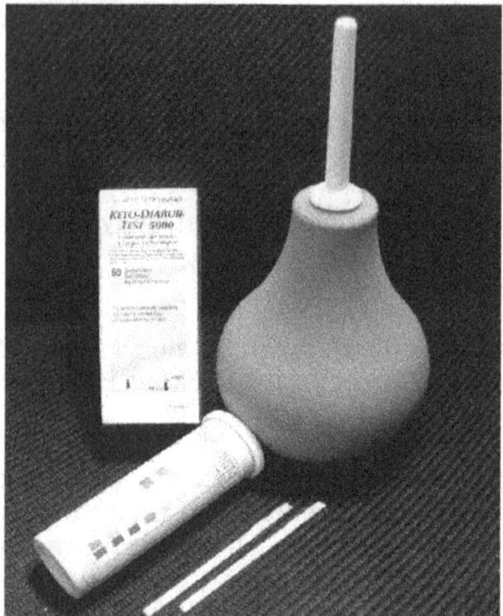

Fig 9.1. Rectal enema syringe and ketone strips are indispensable for long distance traveling to "exotic" countries

Can you summarize the most important points again?

Increased blood glucose, hyperglycemia, can occur even though you are not conscious of any mistakes in your insulin dosages. The reasons for this are usual variations in insulin requirements from day to day, irregular absorption of insulin, particularly delayed-acting insulin, and most often - miscalculations (or forgetting!) of prandial insulin. Thus, blood glucose self-checks *and corrections* are necessary several times a day. If hyperglycemia is short-term and temporary it is insignificant. Do not worry about it. Correct hyperglycemia to the target level with a corresponding bolus of rapid insulin.

If it is obvious that your total daily insulin requirement changed or your insulin dosage algorithms are no longer appropriate for you, then change them. Correcting your FIT algorithms (see the next chapter) is less often necessary than correcting your blood glucose, which is easier.

Really high blood glucose increase requires a double dosage of correctional rapid insulin, as well as blood glucose tests with aggressive corrections as long as the values are not back to normal again. Furthermore, since an acute illness (or vomiting) leads to a dramatic increase of total insulin demand, ketone testing in urine should become a necessity. Ketones in the urine point out the total insulin deficiency, which has to be corrected immediately. Doubling the insulin requirement is common during infections, therefore do not hesitate, raise your insulin dosage immediately.

Liquids or water is very important (including electrolytes), so please drink, several cups or more per hour. Potassium and magnesium solution would be essential, especially if intense blood glucose corrections are needed, since after insulin administration potassium levels in blood diminish.

Ketoacidosis could be avoided even after vomiting and diarrhea. If drinking should become a problem due to vomiting use rectal enema syringe for fluid supply. In fluid depleted state your body will absorb enema quickly from your intestine.

10. Summary of Rules for Algorithm Adjustment

Can I really change my insulin dosage algorithms independently? So, if something goes wrong I can bring my blood glucose to my target level by injecting rapid insulin or by eating carbohydrate-rich foods.

Caution needs to be taken when you change the insulin dosage algorithms that have been effective previously. As long as the average or mean daily blood glucose remains at approximately 120-160 mg/dl, and you have a near-normal hemoglobin A_{1c} value without severe or frequent hypoglycemia, there is absolutely no reason to change your rules of insulin dosage (MBG should include spot measurements *after* meals). Remember that set MBG targets *need* to be somewhat higher (130-200 mg/dl) in people at risk of serious hypoglycemia.

When do I need to change my insulin algorithms?

When your daily insulin requirement changes or when your rules for basal, prandial or correcting insulin dosage are no longer accurate.

How can I identify whether my insulin requirement has changed or when the algorithms are no longer accurate?

By consulting your "balance records". **Never change your rules for insulin dosage without checking your daily balance over the last couple of days (i.e. total daily insulin U/day, ratio of delayed-acting insulin to rapid insulin, MBG, carbs/day).** If the daily MBG exceeds 100-160 mg/dl on a particular day, that is okay. This may have been caused by an upward or downward blip. However, if the value remains over 160 or under 100 mg/dl **on three consecutive days**, this is a signal that your insulin dosage is not adequate. At this point, you need to consider whether a secondary adjustment of your insulin dosage is necessary.

What does "secondary adjustment of the insulin dosage" mean?

We describe the **"primary" adjustment** of insulin dosage as simple corrections for blood glucose with rapid insulin for high blood glucose episodes or with carbohydrates for hypoglycemia. This primary adaptation happens permanently because it can be done after almost all blood glucose measurements. Usually, even with correct insulin replacement on a daily basis in a relatively "normal" lifestyle, most people need to correct every second or third measured blood glucose value. That much of self-monitored values are often outside the target level.

The **"secondary adjustment"** of insulin dosage is not a simple correction of blood glucose but a change in the values for algorithms of the insulin dosage. You can see from your personal balance record whether a secondary adjustment is necessary. Then you need to take the daily insulin requirement into consideration as well as your MBG.

When are changes in my insulin rules necessary?

An **increase in blood glucose** (and thus in your insulin requirement) occurs during:

- Menstrual cycle, 2nd half,
- Acute illness (especially if an infection with fever is present),
- Surgery,
- Puberty,
- Pregnancy,
- Weight gain,
- Chronic stress/school (teachers included☺).

A **decrease in blood glucose** (and thus in your insulin requirement) occurs:

- During physical exercise,
- During weight loss or reduction in food consumption,
- At the end of puberty,
- When giving birth,
- With decreased kidney function,
- With improved insulin sensitivity due to improved glycemic control.

When I realize I need more or less insulin, what do I need to do?

If you do not recognize any particular cause for increase or decrease of insulin need, wait at the most 3 days and, during this time, correct blood glucose with rapid insulin or carbs. Then, ask yourself the question: "Will the changed insulin requirement continue?" If the answer is "yes" (in any case with clearly recognizable causes like after giving birth, when insulin requirements can drop dramatically, or after surgery, when insulin requirement can increase sharply) you will need to change your algorithms.

If I need to correct frequently and I know my insulin requirement changes, how can my rules for insulin dosage be adapted?

This depends, if your total daily dosage changes, adopt your prandial and basal in the same proportions. If numerous corrections for blood glucose with rapid insulin double your daily insulin requirement, you know that you need to double the basal and prandial insulin instead. Another example: you will need to halve your insulin dosage during long intense physical exercise. Your algorithms for basal and prandial insulin need to be reduced by half as well. After a vacation (or a few days after surgery), your daily insulin requirement will bounce back to your "usual standard" and you will need to change your rules for insulin again.

Do I ever need to change single insulin algorithms?

You need to correct single algorithms only when there is no concrete evidence to change your entire daily insulin requirement or when you repeatedly experience the same problems with your current insulin dosage. "Single rules" (algorithms) need to be changed (increased or decreased) no more than 20% at a time.

INTERNATIONAL RESEARCH GROUP
ON FUNCTIONAL INSULIN TREATMENT
Medical University Vienna
kinga.howorka@meduniwien.ac.at
www.diabetesFIT.org

IQNet

Name :..........**Hellen**......................................
Birth date:......................Phone:............
Address:..
E-mail..
Diabetes.since:..................Wt.:.............
FIT since:............ with O injections O pump

I BASAL (fasting insulin): AM ***11 LAN / 2 HUM*** U N late PM ***11 LEVEMIR*** U S PRANDIAL: for 1 carb choice (15g) = ***1,5 HUMALOG*** U U	**Target for correction of aberrant BG values:** Fasting/pre-meal: 100 mg/dl (or ....) After meals: 1h<180 (or.............), 2h< 140 mg/dl **Target range for MBG: from .... to .... mg/dl**

L CORRECTION: 1 U rapid ins. lowers my BG by approx. ***- 40***.......; 1 carb raises my BG by approx. ***+ 60***..........mg/dl
I EXAMPLE: carb choices/CHO: ....
N insulin (U): ....

	TIME	1	2	3	4	5	6 AM	7	8	9	10	11	12	1	2	3	4	5	6	7 PM	8	9	10	11	12	Total Daily Dose	
MON	Basal							*11 LAN*															*10 LEV*			*21*	
	Bolus	*Humalog*						*8*			*2*			*4*					*5*			*3*				*22*	*43*
12.	BG							*146*					*120*					*149*						*107*		MBG	*131*
Oct.	Carbs/CHO							*3*			*1*			*3*					*3*			*2*				*12*	
	Comments																										
TUE	Basal							*11 LAN*															*10 LEV*			*21*	
	Bolus							*4*			*5*			*3*		*2*					*6*				*2*	*22*	*43*
13.	BG							*86*		*~300*						*196*				*166*			*182*			MBG	*186*
Oct.	Carbs/CHO								*2*			*1*		*2*								*3*				*8*	
	Comments	*Sore throat!*																									
WED	Basal							*11 LAN*															*10 LEV*			*21*	
	Bolus							*8*			*3*			*8*					*9*			*8*		*4*		*40*	*61*
14.	BG						*243*						*239*					*320*			*242*			*161*		MBG	*241*
Oct	Carbs/CHO								*1*		*2*				*2*					*½*		*2*			*~1*	*~8*	
	Comments	*Shivering, 100°F Fever*																									
THU	Basal																										
	Bolus								*17*																		
	BG								*387*																	MBG	
	Carbs/CHO																										
	Comments	*Urine glucose++ Ketons++*																									
FRI	Basal																										

Fig. 10.1a: Need to adjust algorithms?

Helen has been having a cold since Tuesday. When (at the latest), and how does she adjust algorithms? Her total daily insulin dose was usually around 40U per day.

Fig. 10.1b: Although several blood glucose corrections have been carried out with short acting insulin since Wednesday, sufficient glycemic control could not be reached: Mean blood glucose (MBG) remained over 160mg/dl. Ketones in the urine suggest a higher insulin dosage is necessary.

Since Wednesday, Helen has been taking antibiotics but the fever has not gone away. With such a serious flu, a continuation of the increased insulin requirement needs to be assumed, so an adjustment of the present algorithms of insulin dosage is necessary! Because the insulin requirement has risen by at least one-third, the basal rate, as well as the dosage for every carbohydrate choice needs to be increased by one-third.

The new dosage would then be approximately:
Basal insulin:
    mornings: 15 U Lantus® / 3 U Humalog®
    late evenings: 15 U Levemir®
Prandial insulin:
    for every CHO: 2 U Humalog®

Notes:
Sometimes, the insulin requirement rises without any apparent reason (e.g., in women sometimes in the 2nd half of the menstrual cycle or before menstruation). In such cases, a secondary adaptation of the insulin dosage (algorithm correction) needs to be considered. At the latest, on the third consecutive day with MBG over 160mg/dl or the other limit chosen.

How can I change single insulin algorithms? In particular, when to change the basal dose?

Signs of an insufficient *(low)* basal insulin dosage are:

- High fasting blood glucose values
- The proportion of delayed acting insulin in the daily insulin requirement is less than 35% of delayed acting insulin
- Repeated ketones in the urine
- Repeated blood glucose corrections that are continuously necessary without any readily apparent "cause".

In this case, an increase in the basal dose needs to be considered (the first step: 10-20%).

Signs of an overly *high* basal insulin dosage are reasons to reduce the basal rate:

- Repeated low fasting blood glucose values (values lower than 90 mg/dl are only occasionally acceptable)
- Overnight spontaneous blood glucose reductions over several consecutive days
- A high percentage of delayed acting insulin (in adults over 50%, in children over 60%) as part of the daily insulin requirement
- The necessity of eating to "keep-up" with the basal insulin: Eating food without prandial insulin or exercise does not lead to hyperglycemia
- Repeated hypoglycemic episodes for no apparent reason
- Weight reduction: your liver serves as a glucose storehouse. When you lose weight, you use this energy supply. When the liver's "storage" (glycogen) is low, your regular basal rate will be too high. Therefore, you need to reduce it over time.

I remember that the fasting insulin dosage needs to be set so that it causes no spontaneous decrease in blood glucose between meals. It, therefore, guarantees fasting values mostly between 90 and 140 mg/dl and it needs to be a maximum of 50% of the total daily insulin dosage, right?

Yes. If short acting insulin equals to much less than two thirds of the total daily insulin, that may also be a sign that the established **prandial algorithms are too low.** This is particularly true if repeated high blood glucose values are measured several hours after eating.

In contrast, prandial algorithms that are **too high** cause hypoglycemia, when it occurs repeatedly and the proportion of rapid acting insulin is more than 60-70% of the total daily insulin requirement. Remember, it is not advisable to adjust the prandial insulin by more than 20% at any given time.

What needs to be considered when changing blood glucose correction algorithms?

The corrective algorithms: change in blood glucose per 1 unit rapid insulin" or change in blood glucose per 1 carb choice relies on the amount of the basal insulin dosage. If a spontaneous decrease or increase in blood glucose results from an overly high or low basal rate, then correcting algorithms will not be accurate. With correct basal replacement, patients who weigh approximately 130 pounds and with a daily insulin requirement of 40 -60 units can expect one unit of rapid insulin to lower their blood glucose by about 40 mg/dl. Basically, the initial algorithm change in blood glucose per 1 unit rapid insulin" can be calculated by the following formula:

1700/ Total Daily Insulin Dose = _________ (round up) mg/dl

Ask your physician if the value applies to your particular case.

The correcting value need to be *increased* in the following situations:

- When the insulin need is decreased (i.e. physical exercise, after pregnancy…)
- At the beginning of FIT, temporarily for safety precautions
- Temporarily in hypoglycemia risk situations
- Permanently in people who are at high risk of hypoglycemia, and in children as well

Correcting algorithm need to be *decreased* when:

- The daily insulin requirement is high (or increasing)
- The person has high body weight.

What else do I need to consider when determining blood glucose correction targets for myself?

Set an adequate correction level. Do not allow printed record sheets to influence "*your*" correction target. Values such as pre-meal target of 100 mg/dl and post-meal lower than 160 mg/dl are only "examples". The correction target level needs to be increased:

- Permanently in people who are at risk of serious hypoglycemia
- Temporarily in all hypoglycemia risk situations(like driving)

Lowering blood glucose correction targets is necessary only during pregnancy or planning for pregnancy. Sometimes a lower target can be achievable in type 2 diabetes.

INTERNATIONAL RESEARCH GROUP ON FUNCTIONAL INSULIN TREATMENT
Medical University Vienna
kinga.howorka@meduniwien.ac.at
www.diabetesFIT.org

Name :..................**Eve**....................................
Birth date:........................Phone:............
Address:...
E-mail...
Diabetes.since:..................Wt.:.............
FIT since:............ with O injections O pump

I BASAL (fasting insulin): AM ..../.........................U **N** PM ....U	**Target for correction of aberrant BG values:** Fasting/pre-meal: 100 mg/dl (or ....)
S PRANDIAL: for 1 carb choice (15g) = ...U **U**	After meals: 1h<180 (or..............), 2h< 140 mg/dl **Target range for MBG: from .... to .... mg/dl**

L CORRECTION: 1 U rapid ins. lowers my BG by approx.- ....; 1 carb raises my BG by approx.+ ....mg/dl

I EXAMPLE: carb choices/CHO: ....
N insulin (U): ....

	TIME	1	2	3	4	5	6 AM	7	8	9	10	11	12	1	2	3	4	5	6	7 PM	8	9	10	11	12	Total Daily Dose	
MON	Basal	Lantus						10												10						20	34
	Bolus	Humalog						6					4							4						14	
27.	BG							174						98						110				87		MBG	
3.	Carbs/CHO								2				3	1							4			0,5		~11	
	Comments																										
TUE	Basal								10															10		20	34
	Bolus								6					3					2					3		14	
28.	BG							198			110			96										111		MBG	
3.	Carbs/CHO										1			3					2			2		3		11	
	Comments																			Jogging							
WED	Basal		Alarm clock!						10																		
	Bolus								8																		
	BG			131					187																	MBG	
	Carbs/CHO									3																	
	Comments																										

Fig. 10.2a: Eve is not happy with her fasting blood glucose values.
Hypos in the night are not likely. Total daily insulin dose is lower than 35 units. What would you do if you were in her place?

- Use Lantus® in the evening and just once a day?
- Use more Lantus® both in the morning and in the evening?
- Use another delayed-acting insulin instead of Lantus® late in the evening?

Fig. 10.2b: Already more than half of the daily insulin dosage (20:34 U) consists of delayed-acting insulin. Another further increase of the basal dosage is therefore not recommendable. Most efficient against high fasting values is probably solution 3: Instead of Lantus® another insulin of NPH type or Levemir® should be injected late before going to sleep (in the same dosage as Lantus®, or even at least in the beginning maybe 10-20% lower). Lantus® in the morning can remain unchanged or perhaps even slightly reduced. In this way not only the high fasting values could be efficiently lowered (see table 4.2) but also the basal insulin dosage could be probably reduced. Experience shows the disadvantage of having two different basal insulin types can be largely neglected, since injection times can be associated to bed- and waking-times.

If you prefer the first solution, i.e. using Lantus® just once a day and only in the evening, please remember: Lantus® is not designed to influence a limited time of the day specifically (e.g. high fasting values), because the effect of this insulin spreads quite equally over a time period of 20-28 hours. The best results with "Lantus® once daily", i.e. adequately low morning values, have been shown with Lantus® before dinner. Here, prandial insulin amount for dinner can be slightly reduced, too. However: total Lantus® dose should never exceed 50% of total daily dose!

Could you summarize this chapter?

The primary adjustments of insulin dosage or exclusive blood glucose corrections are sufficient as long as:

- The average daily blood glucose (including values after eating) remains at 100 (110) -160 mg/dl
- No serious hypoglycemia occurs
- Only half of the daily insulin or less consists of delayed-acting insulin.

Values above the target level should be corrected by rapid insulin. Values below the target area need to be corrected with carbs.

The secondary insulin dosage adjustment or algorithm correction is performed rarely. It is to be performed only when:

- There is an overall change in the total daily insulin dosage recognized by high MBG values and/or many daily blood sugar corrections;
- Your current algorithms for insulin dosage are no longer adequate for you.

Your personal algorithms are to be changed only when you can assume the continuation of the changed insulin need. Increasing or decreasing daily insulin dosage requires algorithms for the basal and prandial insulin components to be adjusted proportionally (at the same ratio), as the average daily insulin requirement.

As a routine, the following criteria need be used to judge the basal insulin dosage:

1. Blood glucose stability during short-term fasting between meals
2. Fasting, blood glucose values
3. The ratio of delayed-acting insulin to rapid insulin in your daily insulin balance.

If average fasting blood glucose values remain above 140 mg/dl despite an increase of the basal insulin dosage (to a maximum of 50% of the total daily insulin requirement), the basal rate needs to be increased as long as (night-time) hypoglycemia has been excluded. NPH and Levemir® administered late in the evening before going to sleep are better than Lantus® for elimination of high fasting values because their duration is shorter. In a "combined" basal replacement NPH or Levemir are given before going to sleep while Lantus® is kept in the morning. Insulin pumps can be pre-programmed for increased delivery in the morning hours when needed.

Basal insulin (long acting component) should not spontaneously decrease or increase blood glucose between meals. The dosage of long acting insulin needs to be as low as possible. So low in fact those carbohydrates eaten will increase blood glucose and so low that fasting blood glucose values are only rarely under 90 mg/dl.

Bolus rapid insulin is used for "morning mound" and for correctional and metabolic purposes. More than half of the daily insulin dose is to be rapid insulin for prandial and correctional needs. Correct dosage of prandial insulin guarantees near-normoglycemia (blood glucose values within the target level) several hours after a meal. When changing single insulin algorithms, do not adjust more than 20% at the time.

11. Physical Exercise

I already know that physical exercise lowers blood glucose if I do not eat something to keep up with my basal or reduce my insulin dosage. Which is better for exercise: eating more or reducing the insulin dosage?

That's not so easy to answer! Physical exercise does often leads to a drop in blood glucose in those treated with insulin. However, there is an exception in the case of absolute insulin deficiency, as in type 1 diabetes. Blood glucose falls during exercise with mild hyperglycemia, but it may rise when you exercise with very high blood glucose levels. If you inject too little insulin for any reason, or if you detect acetone in your urine, then physical exercise may increase your blood glucose and ketones. Until you have corrected the insulin deficiency, you will need to refrain from doing strenuous physical activity.

Now I can remedy any high blood glucoses by using rapid insulin to correct. Please tell me more about the effect of exercise on glucose and insulin.

In someone without diabetes, insulin production during exercise decreases and glucose production increases. So when insulin is available, it increases absorption of glucose by the muscles. However, this effect of exercise, does not happen with total insulin deficiency. Exercise may actually increase your glucose level if there is not enough insulin to "unlock" the cells. This can cause hyperglycemia due to the stress hormones increasing glucose production by the liver.

INTERNATIONAL RESEARCH GROUP
ON FUNCTIONAL INSULIN TREATMENT
Medical University Vienna
kinga.howorka@meduniwien.ac.at
www.diabetesFIT.org

Name :..............**Peter D**..........................
Birth date:.......................Phone:............
Address:..
E-mail..
Diabetes.since:..................Wt.:..............
FIT since:............ with X injections O pump

I BASAL (fasting insulin): AM ***13 LEV / 4 NovoLog*** U **N** PM ***11 LEV***...................... U **S** PRANDIAL: for 1 carb choice (15g) = ***1.2 NovoLog*** U **U**	**Target for correction of aberrant BG values:** Fasting/pre-meal: 100 mg/dl (or ....) After meals: 1h<180 (or..............), 2h< 140 mg/dl **Target range for MBG: from .... to .... mg/dl**

L CORRECTION: 1 U rapid ins. lowers my BG by approx.- **30**......; 1 carb raises my BG by approx.+ **50**..........mg/dl

I EXAMPLE: carb choices/CHO: ....
N insulin (U): ....

	TIME	1	2	3	4	5	6	7	8	9	10	11	12	1	2	3	4	5	6	7	8	9	10	11	12	Total Daily Dose	
							AM													PM							
MON	Basal	*Levemir*						*13*													*Levemir*				*10*	*23*	
	Bolus	*NovoLog*						*10*			*4*			*5*									*3*			*22*	*45*
10	BG							*146*								*110*					*62*			*121*		MBG	*110*
July	Carbs/CHO							*~3*			*3*			*4*			*2*				*1*		*4*			*17*	
	Comments																	*Football*									
TUE	Basal	*Levemir*						*13*													*Levemir*			*11*		*24*	
	Bolus							*8*			*5*					*6*				*6*			*4*		*1*	*30*	*54*
11	BG							*97*			*147*							*99*				*121*		*184*		MBG	*130*
July	Carbs/CHO								*3*		*4*					*5*				*5*			*3*			*20*	
	Comments																										
WED	Basal	*Levemir*				*13*															*Levemir*			*10*		*23*	
	Bolus					*4*								*3*					*4*							*11*	*34*
12	BG	*from today*				*120*					*HYPO!*			*~130*				*87*						*67*		MBG	*101*
July	Carbs/CHO	*summer job:*					*2*		*1*		*2*		*5*			*2*			*4*		*K.O.*			*2*		*18*	
	Comments						*Conveyor belt -- 8 hours! daily*																				
THU	Basal						*8*														*Levemir*			*7*		*15*	
	Bolus	*secondary*					*3*							*3*					*4*			*6*				*16*	*31*
13	BG	*adjustment!*					*68*						*121*						*79*			*110*				MBG	*94*
July	Carbs/CHO						*3*	*1*			*1*			*4*					*5*			*6*	*2*			*22*	
	Comments								*Conveyor belt*																		
FRI	Basal						*8*																				
	Bolus						*4*																				
	BG					*134*																				MBG	
	Carbs/CHO						*3*																				
	Comments																										

Fig. 11.1a: Peter D, a student of public health, reacts differently to a short-term physical exercise (Monday football match) than to a long-term, continuous, intense physical activity at his summer job (on a conveyor belt assembly line) which started on Wednesday. Initially, without any relevant exercise, his daily insulin dose was below 50U/day.

Which precautions were taken?

(1) At the football match on Monday?

(2) At work on the conveyor belt assembly line from Wednesday to Thursday?

Figure 11.1b:

(1) Peter's reaction to a short term strenuous exercise

On Monday, when playing football (short-term physical exercise), extra carbohydrates were consumed (1-2 carbs per 1 hour of exercise). Before and after the game, blood glucose was measured. At dinner, only a very small dosage of insulin was administered for safety reasons (because the effect of physical activity can last much longer than the activity itself). Levemir® basal was also reduced by 10%.

(2) Peter's reaction to a continuous, strenuous, summer job.

From Wednesday onward, Peter has been working on a conveyor belt (as a summer job). But with so much activity, the insulin dosage is now too high. Despite "eating to keep-up with the basal rate", hypoglycemia occurred at 10 o'clock. Because Peter could not predict that the physical work would be so intense on the first day of his job, which was now planned for a longer time, the guidelines for a short-term activity were applied (1-2 carbs per 1 hour of exercise). In the evening, one look at the daily balance – total daily dose and food consumption, and mean daily blood glucose – provides a clear picture of what was needed. Despite a high carbohydrate intake, the daily insulin requirement was reduced by about one third. Because Peter planned his job for a longer period, he adjusted his algorithms according to the change in insulin requirement: not only the basal (from 24 to 15 U/day) but also the prandial (from 1.2 to 0.8 U per carb) dosage needed to be reduced by about one third.

Caution! With falling insulin need due to exercise, the correction algorithm "1U of NovoLog® will lower my blood glucose" will increase as well, e.g. from -30 mg/dl to -50 mg/dl.

You said that during physical exercise insulin production drops in people without diabetes. What happens in people with diabetes?

Insulin production does drop in people without diabetes and this can induce some problems in people with diabetes treatment. For example, if you decide to play football and you have already injected your long-acting insulin, it's too late to reduce your basal insulin. Now, your insulin rate, unlike that of a person without diabetes cannot be lowered further for this football match.

So, would I need to eat something?

Yes. During short-term, sporadic, unplanned physical activity, you will need to eat carbohydrates without prandial insulin.

How many carbohydrates are appropriate?

That depends on your current insulin level and on the duration and intensity of the exercise. Think about when and how much rapid insulin you injected. Approximately 2-3 carbohydrate choices for each hour of exercise are usually needed to be eaten (without prandial insulin) to keep up with the basal insulin and to stabilize blood glucose during intense physical activity. You need to measure your blood glucose before and after exercise to confirm that your precautions were correct during short-term physical activity such as tennis or window cleaning. Moreover, your blood glucose may additionally drop many hours after physical activity because your insulin requirement is reduced. This is due to enhanced *post exercise muscle refuel* with glucose. After every strenuous physical exercise, it would be appropriate to reduce your basis for the evening (by approx. 20%).

Do I need to reduce my insulin dosage for long-term pre-planned physical activity like running a marathon?

Yes, because exercise usually has an effect similar to insulin and will usually lower your blood glucose. Hence, the total daily insulin requirement decreases for people with diabetes. Vigorous, long-lasting physical activity may reduce your insulin requirement and your Total Daily Dose by one-half or more.

Do I also need to decrease my basal and prandial insulin by one-half according to the FIT rules for insulin algorithm adjustment?

Yes. It is not possible to know how much insulin you will need every day during a particular activity. You can only estimate how high your daily insulin requirement will be when you pick strawberries for a couple of hours or go cross-country skiing for a whole day for example. On the first day of long-term, planned physical activity simply apply the rules for short-term, unplanned activity. Work approximately 1-2 carbohydrate choices per hour of activity "into the basal rate" without prandial insulin. Continue to monitor your blood glucose, and calculate your new daily insulin requirement in the evening. Then, you need to make decisions concerning future algorithm adjustments. Reduce all algorithms by the same percentage that you have decreased your average daily insulin requirement. Don't forget to carry out another algorithm adjustment when your exercise or vacation is over.

Note: intensive exercise with insufficient decrease of insulin dosage after the exercise is one of the most common causes of hypoglycemia! Reduce your insulin to keep up with post-exercise muscle refuel.

Will you summarize briefly:

Physical exercise has an effect "similar to insulin". As long as you still have some insulin in your body, cells are "unlocked" to absorb glucose. You may use additional carbohydrates or less insulin than usual during exercise to manage diabetes successfully.

- During **short term**, unplanned physical activity, consuming carbohydrates without prandial insulin is important. One to two (or at the most, three) carbohydrate choices are usually necessary for each hour of strenuous physical exercise.
- During **long term,** planned physical activity, it is better to reduce the basal and the prandial insulin dosages proportionally, i.e. about 30-50%. After taking into account your new total daily dose, it is necessary to adapt your algorithms for basal and also prandial needs.

Check your blood glucose more frequently to ensure that your decisions concerning the dosage were adequate to maintain good glycemic control and to prevent hypoglycemia after exercise. A suddenly started very intensive exercise sometimes can induce a paradox blood glucose increase (increased glucose production in the liver, stress hormones). Correct slightly and with caution! Be aware of post exercise muscle refuel leading to spontaneous drop of blood glucose.

12. Special Situations and Adventures

If I can bring my blood glucose into its target area at any time, it must be impossible to get out of control! Are there really any "special situations" that apply to FIT?

The worst thing that could happen is that you could lose your ability to "control" or "regulate" your blood glucose close to the near normal range. This could happen if an illness takes control over your body and makes you captivated, meaning you cannot take care of yourself. Ask or cry for help, otherwise you are risking your life. Ground principle: State or even write down your impairments and disabilities so you can be properly treated.

What could happen if I ever suddenly lost my self-monitoring equipment and insulin?

Clearly, this would not be classical situation you need to avoid. Be prepared! Keep a spare kit!

But what could I do on trips abroad if I was mugged and my "minimal equipment" got stolen?

Try to replace your "minimum kit" necessary for FIT. "Mini-kit" needs to be carried with you at all times, so it is important that it is replaced immediately.

Wherever you go, "Mini-Kit" for FIT needs to be at your disposal at all times

But, other countries have totally different insulin, don't they?

For fasting insulin needs, you can inject any delayed-acting insulin without excessive worry. If at all possible, avoid using pre-mixed insulin. Pre-mixed insulin is a ready mixture of delayed-acting (NPH) and rapid-acting insulin. Take delayed-acting insulin for your basal replacement, any type you can get, animal or human. In this emergency situation, origin is not so important. For the beginning, use the same dosages as before for your fasting insulin, i.e. similar portions twice a day, in the morning and before bed. But, be aware of peaks and durations of NPH, Levemir® or Lantus® insulin. Zinc formulations such as Lente or Ultralente type of insulin disappeared from the market almost all over the world.

What if I cannot get any delayed-acting insulin?

In this emergency, you can substitute your basal with multiple daily injections of rapid-acting insulin. For this purpose you need to calculate approximately how many insulin units you need per hour. Let's assume you need 24 units of fasting insulin every 24 hours. You inject 12 units of NPH insulin in the morning and 12 late in the evening. You need about 1 unit of insulin per hour of fasting. If only regular insulin is available, you still need 1 unit of insulin per hour even if you don't eat anything. You must inject every 4-6 hours because regular insulin works for only that amount of time. For example, inject approximately 6 units of regular insulin 4 times a day (6 units x 4 injections = 24units) to cover your fasting needs. With rapid acting insulin analogs, even more frequent injections are necessary. If you want to eat something, use rapid insulin for the meal, just as you would do usually.

How can I manage without blood glucose self-monitoring?

Try to obtain self-monitoring equipment as quickly as possible. Keep spare strips for visual estimation by comparison with a color scale. Another possibility in this situation is to use urinary glucose self-measurements temporarily. The object, here, would be to try to maintain negative levels of glucose in your urine.

General recommendation: As each meter disables one day all of a sudden, use visual Betachek (www.betachek.com) strips for such emergencies at home. Cheap and easy to use.

But what if I can't even get urine glucose strips?

Then you have only your symptoms or how you feel to monitor your blood glucose level, which is not ideal. You can try injecting 1-2 units of rapid insulin or to exercise to determine whether hypoglycemia occurs. Also, you need to watch for signs of hyperglycemia, such as frequent urination, dehydration, and so forth. In any case, you are at risk if you do not have self-monitoring equipment and insulin. You need to obtain the necessary equipment and/or seek qualified medical assistance, as soon as possible.

It would be really bad if I got sick on top of everything else. I am not sure if I could cope alone.

It is not advisable to be without self-monitoring equipment and your recommended insulin. If you are so sick that you cannot treat your own diabetes, you need to find someone and tell them your needs immediately. State explicitly, that you are not able to take care of yourself!

Does this mean I would have to go into the hospital?

Possibly —if no other solution can be found. Just because you are hospitalized, but otherwise in good health, does not mean you need to give up self-monitoring and self-treatment. Only when you are critically ill or unable to carry out FIT, only then does someone else need to do this for you. When you are feeling better, continue again your self-treatment in cooperation with the physicians overseeing you in the hospital. Show them your log-book for insulin self-dosage. If necessary, suggest a secondary algorithm adjustment. This is particularly important when you foresee an increase in insulin requirements. Increases are common after undergoing surgery, bone fractures, etc.

I have a friend who has diabetes. He had minor surgery and was told that five blood glucose self-measurements were too much. The doctor thought he should not bother with "so many" shots. He was advised to test before main meals at most, to inject pre-mixed insulin twice a day and do nothing else.

FIT is possible only when a person has had relevant training. Few people are able to take care of themselves immediately after surgery or able to correctly carry out insulin replacement. Physicians are unaccustomed to and, sometimes, skeptical toward people with diabetes who treat themselves as effectively as our patients routinely do. Showing your log-book and discussing your treatment program will encourage the physicians to support your approach to diabetes management. You can convince your doctors that you are capable of taking responsibility for yourself. If your medical team does not support your self-monitoring, you need to speak to the head of the department or the nurse in charge and explain why self-monitoring is necessary. If you are still not understood you need to seek other medical partners in the future.

If this is a planned hospital admission for "elective" surgery, you can avoid all these problems. Before going into hospital, ask for information on the surgical procedure required and its options and explain your diabetes treatment to someone who will be looking after you while you are in hospital.

I used to have to go to the Emergency Room for intravenous (IV) fluids when I had ketones and became dehydrated when I had the flu. I hope this will not happen anymore.

If you have a cold or the flu you can expect a general increase in your insulin requirement. Your insulin requirement could double in situations like this. You can now predict how much your insulin requirement will increase but you will need to be cautious.

With FIT, at least initially, I have to constantly correct my blood glucose using rapid-acting insulin (primary insulin dosage adjustment) anyhow.

However, even more important is for you to carry out the secondary adjustment of insulin dosage by changing algorithms when several consecutive corrections of high blood glucose levels confirm that your insulin requirements have really changed. Remember: With very high blood glucose values (e.g. over 400 mg/dl), double the amount of short-acting insulin for correction! When your increased insulin need is over, you will need to reduce your algorithms, step by step, as your insulin need decreases.

Ketone strips are essential and indispensible for testing urine while ill

Apart from the daily mean blood glucose (MBG) and the average total daily insulin dose, what are the indicators for adjusting my algorithms or carrying out a secondary adjustment of my insulin dosage?

One important indicator is ketones excretion in the urine. If you can rule out forgetting to inject basal insulin, a positive acetone test and high blood glucose values are evidence of an absolute lack of insulin! Danger! The insulin injected is not enough – it's too low. In this case, you need to increase your insulin dosage and re-adjust your algorithms. Theoretically, a state of hunger, like fasting or very low calorie diet, can cause ketones. However, in this case blood glucose values would be low –perhaps below 85 mg/dl –even with the ketones present in the urine. If you have ketones with blood glucose below the "healthy" blood glucose target at around 90 mg/dl, you do not need extra insulin.

If I am suddenly ill, do I need to test my urine for acetone?

Yes, definitely. As mentioned earlier, insulin requirements increase with fever and in other comparable illnesses, especially, if high blood glucose values are present.

I am so relieved that I no longer need to force food into my system when I have the flu and vomit!

Never skip your basal injection or testing!

Even with vomiting, during an illness you may expect a considerable increase in insulin need. Never skip your basal injection or testing! You might need to increase basal insulin and make a corresponding blood glucose correction, if necessary. Test more often and correct.

Can I safely diet and lose weight now?

If you want to lose weight, you need to eat fewer calories than your body needs for energy. In this case, it is very important to reduce your basal rate as well. As you already know, when you fast or diet, your liver glucose "reserves" deplete, and it produces less glucose. So, your previous basal rate will be too high now. To reduce it, you need to consider your total daily insulin consumption. Remember that no more than about 40-50 % of your daily insulin requirement should be allotted to delayed-acting basal insulin. I would recommend that you speak to your diabetes physician if you would like to lose weight. Instead of going on crash diets, a safer and more effective option, in the long-term, is healthy eating, combined with exercise (about 30 minutes per day would be appropriate) and lifestyle changes.

If I want to lose weight, it's probably good to reduce my fat intake, since fat contains more calories than carbohydrate or protein.

That's right. Daily endurance exercise is even more important as it improves insulin action. It also helps if you watch and even count your daily calories intake. For some people, making simple behavior changes works well, such as drinking water and eating more fruit and vegetables instead of rich, fatty foods.

Are there any special precautions to take if I am invited for dinner and have no idea what type of food will be served?

You need to take your "minimal" FIT equipment with you. Test your blood glucose, see what food appears to you, and decide on the dose accordingly. In "special" situations, such as "dining out", there is an important guideline to follow. Disregard the dictum "*Less is more*". In your case, "*more*" is better than "*less*" when it comes to self-monitoring.

That's logical. I need to know my blood glucose level to calculate the appropriate insulin dose. Still, self-monitoring is not always easy to do. How can I measure my blood glucose 4-5 times a day if I have important duties to perform, like taking exams, working in a factory or driving a car?

Unfortunately, frequent self-monitoring is necessary if you want a flexible lifestyle. You need to compensate for your lack of insulin effectively. You can choose the traditional approach involving conventional insulin treatment where you inject twice/three times daily, eat regular prescribed meal plans at scheduled intervals and obtain your insulin prescription from your physician routinely. However, FIT is a much more effective treatment, offering the benefits of a more flexible lifestyle. Self-monitoring does not demand very much time. You need less than a minute for each measurement. You will probably spend 5 minutes, at most, each day carrying out this important monitoring. These 5 minutes will give you the freedom to eat what, when and how you wish and all the benefits of a flexible approach to diabetes management.

I'm sure I'll manage, but my irregular lifestyle certainly poses difficulties. The problem is that I would like to sleep longer on weekends and rest on Saturday and Sunday. During the week, however, I have to do strenuous physical work.

These are two completely separate situations that can be managed effectively using FIT.

So, if I want to wake up at different times do I inject delayed-acting insulin only once a day late at night instead of twice a day?

Injecting delayed-acting insulin only once a day for the basal needs has too many disadvantages. To compare these with the advantages of taking delayed-acting insulin once or twice a day, see Figure 12.1.

One Injection Daily	Two Injections Daily
Advantages	
• Only 1 injection per day as basal	• Injection time points more variable ("overlapping") • Less fluctuations/More steady insulin absorption • Ease of insulin adjustment for exercise (e.g. nighttime reduction for post-exercise muscle refuel) • "Combined" basal (Lantus® in the morning, Levemir® or NPH late before sleep for as high fasting values as possible • Even if injection forgotten, no DKA
Disadvantages	
• Fixed injection times • Bigger injection volume (causing sometimes more painful injections) • Less stable basal rate with more pronounced peaks • Higher hypoglycemia risk • More fluctuations due to more irregular absorption • Easy development of DKA with even only one injection skipped	• Two injections

© Howorka, Insulin-dependent?, 2009

Fig. 12.1. Advantages and disadvantages of delayed-acting insulin used 1 or 2 times daily

OK, I got it. Taking delayed-acting insulin twice a day is more beneficial. Are there any other things I can do to avoid hyperglycemia or treat the 'dawn phenomenon' in the morning if I sleep-in?

If you wake up at different times it is advisable to inject intermediary insulin late at night (like NPH or Levemir®), instead of long-acting insulin Lantus®. In the morning, doses of Lantus® can remain the same. I recommend that you check your fasting blood glucose (as always) and correct with insulin (as always), if your blood glucose increases. Try not to sleep till midday. Maybe you'll manage to inject the morning delayed acting and 'morning mound' rapid acting basal insulin in a timely fashion at least. Don't forget that your liver produces more glucose in the morning whether you are prepared for it or not.

What do I need to do when my physical activity varies a lot from day to day?

Record your experiences. Continue to carry out your usual blood glucose corrections. Secondary adjustment of your insulin dosage is essential when physical activity differs under these varying conditions. For example, a 19-year-old athlete doubled her fasting and prandial algorithms according to her daily insulin requirement on days without physical activity. Her daily requirements varied from approximately 30-40 units every 24 hours on sports days to around 80 units on free days.

How can I handle time differences when I travel?

On an outward flight, the day will be "shortened" by approximately 6 hours if you travel from West to East. The day will be longer when flying from East to West. If you use an insulin pump, you don't even need to think about the time difference. The basal rate does not change because of the time differences with an insulin pump because it delivers a constant basal rate. If you use injections on your flight from West to East, the basal rate may be too high because of consecutive injections of basal insulin in shorter intervals. To prevent that from happening, reduce the dose of delayed-acting insulin to maybe up to 50% of the usual amount before your flight. In contrast, the basal rate might be too low when flying from East to West, if you "extend" the interval between consecutive injections. Continue to monitor your blood glucose as usual and correct blood glucose increases.

I think I am prepared to cope with just about anything now☺!

In summary, it is important to continue with FIT principles in "special situations" so that the following components are independently replaced:

- Fasting or basal insulin;
- Prandial insulin for meals;
- Correcting high/low blood sugar values, or more specifically, glycemic control consisting of self-monitoring and blood glucose corrections with correction bolus or carbs.

Even in the worst case scenarios you can learn from your experience how to prevent future mistakes, and "improve" future control. If you are no longer in a position to independently continue your treatment immediately communicate this to others. In this case, your treatment needs to be carried out by someone else, family or perhaps your diabetes physician.

In the state of acute illness you can expect a general increase in your insulin requirement. Once again, never skip your basal insulin. Not even if you have to vomit.

13. Insulin Pumps

I have always wanted to try an insulin pump, just for a short time, to see how it works for me. What depresses me is the prospect of always wearing the pump on my body. Can I take a pump off, at least for short periods?

Between FIT with a pump or FIT with injections, there are no real differences. However, if you want to remove the pump, consider that you will have very little insulin depots under your skin, as the pumps provide only rapid acting insulin. If you remove the pump for a few hours only you can use rapid insulin to cover your basal insulin requirements. Inject as many units per hour as the pump would deliver for fasting requirements after taking it off. Here's an example: with a basal rate of 24 insulin units (U) per 24 hours you will need approximately 1 U per hour. If your basal rate is more complex and is higher or lower during defined time periods, take this into account as well.

In the past mentioned basal doses of 1 U per hour, would that mean that I'd have to inject 3 U of short acting insulin if I were planning on taking the pump off for three hours?

Exactly! Concerning the basal rate and switching off the pump: the regular insulin would have to be injected all 4 to 5 hours, short acting analogs would have to be injected at least all 3 hours. If you want to eat, you would have to inject insulin according to your food intake and the same goes for the corrections. If you switch off the pump for longer than 6 to 8 hours i.e. overnight or during the weekend, you might as well continue with FIT, replacing the basal e.g. with an NPH insulin. All other principles remain the same. Please take into consideration that your insulin intake is reduced by using exclusively the short acting insulin in the pump by approx. 10-20% during 24 hours, meaning that your overall insulin intake with the pump will be less than when injecting it with pen.

What advantage will I gain by using the pump?

FIT-treatment, if that hasn't happened until now. Nowadays, insulin pumps are sometimes used as tools *instead* of providing a good FIT-education in various places. There the pump is usually the easiest way how to achieve FIT. If there's no other option, then why not? In some countries, technical support and pumps are even financially refunded (health care reimbursement). Sometimes, it's even easier to receive a pump than to come to FIT. The ethical aspect of such an approach we shall leave with "no comment"...

An additional advantage for the pump therapy would be reducing of the annoying and unpleasant hypoglycemic symptoms. Especially for the small group of patients who have a supranormal hypoglycemic awareness (here, any kind of blood glucose lowering would probably have the same effect, but CSII seems to be even more efficient).

The most important pump advantage is that the pump allows you to shape your basal rate individually. This allows better fasting glucose values.

If one is well aware of the delayed acting insulin, one can have a "combined" basal replacement – i.e. in the morning Lantus® combined with Levemir® or NPH late in the evening – and still achieve good fasting glucose values.

Very well, but even if you don't know this principle, you will obviously be highly attracted by the "pump myth". And I have to confess, the pump technology is becoming more convincing nowadays. Please do take a look at the pump table.

Except the well shaped basal rate, what other advantages would I get with the pump?

Sometimes a better glycemic control can be achieved. Yet you should be careful: even multinational studies which have been financed by pump manufacturing companies show only a decrease of 0,2% HbA_{1c}.

Alright! I know very well that the pump only does that what I do with it. I do not expect any wonders!

The quality of your treatment with the pump is dependent on similar factors just like your therapy using injections, like:

1. How often you measure and correct your BG (often at least 5 to 6 measurements for unstable type 1 diabetes are required).
2. How do you shape your basal replacements? The pump offers here the ultimate advantage: during the night it could deliver less insulin, on the other hand it's possible to deliver more during the morning or before waking up to compensate the dawn phenomenon. The similar effect one can often only be achieved with two different insulin types combined for the basis as described above.
3. How you adapt your insulin delivery according to your meals. Pumps are capable of delivering a "prolonged" or "combined" bolus (beginning with a normal bolus, then a prolonged one).

In order to match the bolus to the food with the injection therapy, you can enhance the insulin absorption with rapid insulin analogs, through intramuscular injection or by improving blood circulation at the injection site by massaging it. To achieve the slow absorption you simply inject regular insulin subcutaneously – for a fatty pizza for instance.

Why does the continuous BG monitoring combined with the insulin pump still not exist? It would do all the work by itself?

The continuous monitoring on its own, not necessarily combined with the pump e.g. System Guardian RT from Medtronic, Freestyle Navigator von Abbott or Seven System from DexCom – will hopefully fulfill many dreams (see, e.g.: www.childrenwithdiabetes.com/continuous.htm).
However, the latest experience has proven that due to technical reasons the patient will still have to rely on him-/herself – and still probably for quite a long time – in estimating his/her insulin dose. If the patient doesn't immediately react upon the glucose level (i.e.: correct instantly), then the treatment results are stunningly miserable. In summary, in either case: for continuous as well as for conventional intermittent blood glucose measurement, corrections are crucial.

I'm correcting the whole time anyway!!

Yes indeed, and this is correct. Our statistical evaluation has proved that type 1 diabetic patients have to correct statistically after every second BG measurement. Well, only for those who haven't successfully completed their FIT education, interesting news exist:

1. Software integration of pump with BG measuring device (look at pump table); among others:

- Dana DiabeCare IISG und Dana blood glucose meter,
- Deltec Cozmo und CoZmonitor blood glucose meter (Abbott, earlier: Therasense),

- Medtronic 522/722 and Paradigm real – time continuous blood glucose monitoring,
- OmniPod und CGM-System-Seven von DexCom (in progress at the moment).

2. Computerized calculation for bolus correction. The IOB ("Insulin-on-board", IOB-Feature) takes overlapping of insulin boluses into account. If not "computerized", a similar function with FIT is the increased postprandial correction target e.g. 160 to 180 mg/dl... (IOB available e.g. in pump models Paradigm 512/712, Animas 1200, and Cozmo among others)

Well... I'm still quite insecure about the pump...

I have summarized for you the advantages versus the disadvantages either when using the pump or the insulin injection therapy as often perceived by the users. Details as well as extra information from convinced pump fans you will find e.g. on the website www.insulin-pumpers.org. You will also find further mentioned reference websites at the end of this chapter.

Pumps are becoming more and more popular. The great supply and variety of catheters and improvement of technology are of great importance for further dissemination of pump treatment. This is relevant for catheter disposal in order to taking off the pump for sex, sports or showering.

Made by	Animas	Deltec	Roche/ Disetronic	Disetronic	MiniMed	Sooil	Insulet
www.	animascorp.com jnj.com	deltec.com	roche.com	disetronic.com	minimed.com medtronic.com	sooil.com	insulet.com
Model	IR-1200	CoZmo	Accu-Chek Spirit	D-Tron Plus	Paradigm 512 / 712	Dana DiabeCareII	OmniPod
View							
Size [mm]	89 x 56 x 18	81 x 46 x 23	81 x 55 x 20	105 x 48 x 21	512: 76 x 51 x 20 712: 94 x 51 x 20	75 x 45 x 19	pod: 41x61x18 PDM: 66x110x26
Weight [g]	88	94	79, 110/136 w.battery, full reservoir + infusion set	125	512: 107 712: 108	51	pod: 34 (full reservoir) PDM: 113 (w Battery)
Reservoir Size	200 U Plastic 2ml	300 U Plastic 3ml	315 U Insuman-Infusat or 3.15 ml vial	300 U pre-filled cartridge Humalog or 3ml vial	176 U / 300 U plastic (1.76 / 3 ml)	300 U plastic, 3ml vial	pod-Reservoir 200 U
Connection	Luer lock	Luer lock	Luer lock	Luer lock	Proprietary	Proprietary	Built-in
Colours	blue, silver, black	blue, black, atomic purple	blue, with 30 pump "skins" colours/styles	black, blue	clear, blue, smoke, purple	black, gray, blue, ivory	white

Fig.13.1. Overview/Comparison of Insulin Pumps

Made by		Animas	Deltec	Roche/ Disetronic	Disetronic	MiniMed	Sooil	Insulet
www.		animascorp.com jnj.com	deltec.com	roche.com	disetronic.com	minimed.com medtronic.com	sooil.com	insulet.com
Model		IR-1200	CoZmo	Accu-Chek Spirit	D-Tron Plus	Paradigm 512 / 712	Dana DiabeCareII	OmniPod
BASALRATE (BR)	Incre-ment	0.025 U	0.05 U	0.1 U, from 0.1 to 25 U/h	0.1 U	0.05 U	0.1 U	0.05 U/h, up to 30 U/h
	Basal interval	30 min	30 min	60 min	60 min	30 min	60 min	3 min
	Total basals	12/day	48/day	24/day	24/day	48/day	24/day	48/day
	Basal profiles	4	4	5	2	3	1	7
	Temporary basal	-90% to +200% in increments of 10% for 0.5 to 24 h	in 0.05 U increments, or -10% to +150% in increments of 5% for 0.5 to 72 h	in 10% increments from 0% to 200%, for 15 min to 24 h	in 10% increments from 0% to 200%, for 60 min to 24 h	increment of +/–0.1 U for 0.5 to 24 h or as % of current BR	in increments of 25% between - 100% to +100% for 1 to 12 h	% or U/h for 1 to 12 h, in 30 min increments
	Insulin-On-Board	Yes / Yes	Yes / Yes	No / No	No / No	Yes / Yes (for 522/722)	No / Yes	Yes / Yes
BOLUS	Increment	0.05 visual or audio, 0.1, 1, 5 U audio	0.05, 0.1 visual, 0.05, 0.1/0.5/1/2/5 U visual or audio	0.1, 0.2, 0.5, 1, 2 U	0.5 or 1 U	0.1 visual, 0.5 or 1 U visual or audio	0.1, .05, 1 U	0.05, 0.1, 0.5, 1 U
	Bolus type	standard, extended, combination	standard, extended, combination	quick, extended, multiwave, Scroll	quick, extended, fine	standard, extended, combination (optional wireless/re-mote con-trolled bolus)	standard, extended, combination	standard, extended, combination
	Bolus/ Time	1 or 4 sec	Adjustable, 1-5 min	5 sec	5 sec	30 sec.	13 sec	40 sec.

Made by	Animas	Deltec	Roche/ Disetronic	Disetronic	MiniMed	Sooil	Insulet
www.	animascorp.com jnj.com	deltec.com	roche.com	disetronic.com	minimed.com medtronic.com	sooil.com	insulet.com
Model	IR-1200	CoZmo	Accu-Chek Spirit	D-Tron Plus	Paradigm 512 / 712	Dana DiabeCareII	OmniPod
Battery type	AA Lithium	AAA Alkali	AA Alkali or recharcheable	Proprietary	AAA for pump, A23 for remo-te control	1/2 AA 3.6 V Lithium	AAA x 2 (PDM)
Battery life	6-8 Weeks	4 Weeks	4 Weeks	4 Weeks	3 Weeks	8-12 Weeks	4 Weeks
Memory	Non-volatile: 600 bolus, 270 BR, 120 daily totals, 30 alarms, 60 primes	Non-volatile: 90 days (4000 events) basals, carb boluses, corrections, alarms	Non-volatile: 90 days (4,500 events); also temporary BR–changes	Non-volatile: last 10 boluses, 10 alarms, 7 daily totals; on PC download-able 1400 events (ca 90 days)	Volatile: loss of data can occur; 24 boluses, 7 daily totals, 4000 events	Non-volatile: Last 100 boluses, 100 daily totals, last 100 alarms	90 days (up to 5400 events)
Software/ Download	IR-data transmission ezManager Plus and IR Kit, cradle & software at Animascorp	IR-data transmission CoZmana-ger: program download of last 4,000 events at Deltec	IR-data transmission Pocket Compass with bolus calculator, insulin pump configura-tion software	IR-data transmission Accu-Chek DiaLog, InSight (web) Software	RF data trans-mission; Medtronic CareLink Mgmt System und ParadigmPA L; Model Paradigm 522/722 optional with glucose sensor CGMS Guardian	Wireless data transmission between pump and blood glucose meter by Sooil DanaMagic software	Wireless data transmission between pod and PDM
Alarms	beep, vibration; safety key-lock	beep, vibration; safety key-lock	beep, vibra-tion, combination; safety key-lock	beep, vibration; safety key-lock	beep, vibration; safety key-lock	beep, vibration; safety key-lock	alarm options available
Water?	Waterproof IPX8	Waterproof IPX8	Waterproof IPX8	Water resistant to splashes	Water resistant to splashes	Waterproof	Waterproof IPX8
Warranty	4 years	4 years	4 years	4 years	4 years	4 years	4 years

IR = infrared RF = radio-frequency

Insulin pumps – special product characteristics:

Animas IR 1200

- Bolus manager takes IOB = "insulin on board" into account, thus avoiding interference or overlapping of boluses and suggests the user an insulin dosage according to blood glucose, targets and variable prandial insulin requirement.
- Virtual pump

CoZmo

- Combination with Abbott FreeStyle glucose meter (CoZmonitor) interpreted for the user via pump display,
- May be very variably adapted by the user,
- Bolus manager takes IOB = "insulin on board" into account, thus avoiding interference or overlapping of boluses and suggests the user an insulin dosage according to blood glucose, targets and variable prandial insulin requirement.
- Carb Info database with 600 items.

Accu-Chek Spirit

- Three operating menus according to user's experience and requirements (standard, advanced and custom),
- Software Accu-Chek Pocket Compass for mobile data management on PDA/handheld, basing on blood glucose values and insulin released by pump,
- Two pumps in a set (operation time up to 2 years each, then inspection necessary),
- Optional Carb Info database for PDA.

Accu-Chek D-Tron

- Works with pre-filed Humalog 3ml penfills,
- Accessory software for programming (DiaLog) and/or analysis of data (InSight)
- Two pumps in a set (operation time up to 2 years each, then inspection necessary)

Medtronic MiniMed

- No Carb Info
- Bolus manager takes IOB = "insulin on board" into account, thus avoiding interference or overlapping of boluses and suggests the user an insulin dosage according to blood glucose, targets and variable prandial insulin requirement.
- Further model development (Paradigm 522/722) towards a closed loop system including the CGMS sensor "Guardian"

Dana DiabeCare II

- Optionally a telemetric communication between pump and BG test device by Sooil
- Optionally data transfer to Palm-Pilot PDA or desktop computer, various software („DANA Magic™")

OmniPod

- No Catheter, separate insulin-pod of PDM
- Built-in FreeStyle-Flash glucosemeter
- Only 72 h lifetime of pod...

(In co-operation with Dr. Andreas Thomas, Dia Real/eu-medical GmbH, Dresden. Scale used in individual figures of devices is not homogenous; for size comparisons see data in table)

Source: Respective webpages of products, and/or
www.diabetesnet.com/diabetes-technology/insulin_pump_models.php
www.diabetesnet.com/diabetes-technology/insulin_pump_models_old.php
www.cndm.net/pump/fdapump.php
www.insulin-pumpers.org
www.pharmacytimes.com/Article.cfm?Menu=I&ID=4633

Advantages	Disadvantages
Instead of an average of 6 injections a day, you will have to change catheters only every or every other day, which hurts about as little as a pen injection. Male compassionates with dense fur on their belly excluded (plaster☺).	You will have to wear a little gadget on you all the time. You can always lay it off for a night or a few days and switch to multiple daily injections again. Hopefully you will perceive the pump as an integral part of your body. However, the catheter needs to be changed, and reservoir filled up immediately, whenever the insulin tank is emptied... which tends to happen at the most impossible moments... ☹
You can easily hide your insulin dosing. The times back then when you had to look for some bare skin in winter at a take-away or even, fresh up' in the bathroom in a fancy restaurant are long over (nowadays girls inject their insulin through their nylon hold-ups☺). All you have to do with a pump is to press a button through the textile of your pocket, and the pump will inform you silently (e.g. by vibration) of the amount of U to be confirmed before it injects the bolus.	Since you only inject rapid acting insulin, the danger of ketoacidosis is highly increased: if there is a malfunction of the pump, a knick in the catheter, or if the catheter slips out without inducing alarm, you will end up with a complete lack of insulin, which in turn results in the accumulation of ketones and metabolic acidosis. The probability of a ketoacidosis is much higher by far in pumps than with multiple daily injections. That's why periodic blood glucose measurements are even more important.
The pump allows to specifically target the circadian variation in insulin requirements – the dawn and dusk phenomena. Thus, the probability of night time hypo is lower (possible exception: alcohol – be cautious!). Meals can be postponed or left out at your own discretion. On the weekends, you can sleep late (almost as good as "combined" basal rate).	If the catheter insertion gets inflamed it can leave a scar, due to wearing it for a longer time. Such scars only seldom increase your attractiveness☺.
If you tend to be under a lot of stress at work or just having a lazy day off: the pump allows you to shape your basal rate individually. This allows better fasting glucose values. Reducing the basal rate prevents hypoglycemia after an exercise (post exercise muscle refuel).	Some patients can experience pronounced psychological discomfort. Even if pump therapy is consciously preferred, it may cause anxiety or anguish, being less attractive for the opposite sex, not being able to live like a "normal" person. Injections cannot be seen as much or as vivid as the catheter inserted permanently on your abdomen. For this reason perhaps can the pump cause some problems in sexuality.
Pronounced fluctuations of the BG, occurring with inadequate basal rate (i.e. "only" 2x daily NPH–insulin or "only" 1x Lantus) appear less often with pumps, continuous delivery. Especially for the small group of patients who have a supra-normal hypoglycemic awareness: there is a reduction of the annoying and unpleasant hypoglycemic symptoms.	Statistical decrease of HbA_{1c} for only 0.2% is for a lot of patients not much of an argument for enduring the constant "prostheses" sensation as well as annoyance while having sex. So you have to weight out pros and cons, for the use of pump as well as for the risks of diabetic late complications.
In comparison with insulin injections, if pump settings are properly adjusted, you'll need 10-20 % less insulin.	On the other hand, you'll need to change catheters frequently (usually every other day), in the case that additional 5-15 U of insulin are used to fill the catheter before new insertion.

Tab. 13.2.: Summary of potential advantages & disadvantages of insulin pumps

14. Diabetes Late Complications, Associated Diseases and Sex

Will I get late diabetes complications despite FIT?

Probably no severe ones, as long as your blood pressure is as low as possible and your blood sugar all right. And if you succeed the first 20 years of diabetes without developing any late complications, then it is much less likely for them to occur.

So how come nobody hasn't told me anything earlier?

Over the past few years new evidence has been brought to light. The great studies of DCCT (Diabetes Control and Complications Trial, 1993) and the follow up study EDIC (Epidemiology of Diabetes Complications, where only half of the patients were treated almost as well as our own), have confirmed that good glycemic control is advantageous for your health and your prognosis. With additional lowering of blood pressure, you will finally gain even more than with near normoglycemia only which is difficult enough to reach.

Late complications of diabetes (blindness, kidney failure and nerve damage) develop due to a chronic increase in blood glucose. Hyperlipidemia (high blood fat) and hypertension (high blood pressure) considerable enhance their development. Your self-monitoring is what really matters and the consequences you take from the measurements. Each degree of lowering your MBG/HbA1c and of blood pressure reduces the risk of late complications. Contrary to earlier assumptions, the fluctuations of your blood sugar are probably not very relevant as long as they are corrected instantly several times a day. So once again, remember, what makes the real difference are *immediate* corrections.

Each degree of lowering your MBG/HbA1c and of high blood pressure reduces the risk of late complications

Apart from good glycemic control, are there any other measures I could take?

Minimize all other risk factors for vessel damage such as smoking, high blood pressure, and hyperlipidemia (high blood cholesterol). Also, research has proven that heavy alcohol consumption hastens the development of retinal changes (retinopathy) in people with diabetes. So, refrain from that. Once again, when the late complications start, blood pressure becomes more important than your sugar.

Which mechanisms contribute to the occurrence of late complications?

The typical late complications occurring in the eyes, kidneys, heart, and feet are usually due to combined occurrence of the following components:

Microangiopathy: characteristic changes in small blood vessels such as thickening of the capillary walls, vessel closure, and disturbance of vessel permeability. All lead to poor blood circulation. The retina and kidneys are most seriously impaired by this dysfunction. Loss of vision and kidney failure can result.

***Neuropathy*:** damage to the nerves concerns the *sensitive* and *autonomic* (not voluntarily controllable) nervous system that influences the heart, blood vessels and digestion. This can lead to erectile dysfunction and "diabetic foot".

***Macroangiopathy*:** atherosclerosis (changes of the arteries), which occurs frequently, affects people with diabetes at an earlier age more often and more severely than people without diabetes. A narrowing of the large blood vessels, atherosclerosis or macroangiopathy, can finally result in vessel closures, which lead to heart attacks (myocardial infarction), stroke, and window shopper´s disease.

***Infections*:** these occur mainly in people whose glycemia is not sufficiently controlled. Typical infections are of the bladder and kidney, fungal and purulent skin infections.

Can the late complications be reversed?

Unfortunately, it is believed that from a certain point onward, the late complications cannot be reversed. For this reason, attaining near normal glycemia and low blood pressure early in the course of diabetes and keeping them low for as long as possible are important. Good news: we could induce and observe for a longer time several cases of regression of serious retinopathy with optimal treatment.

How are the eyes affected in diabetes?

The changes in the retina (retinopathy), visible with an ophthalmoscope, mostly occur only after 7-15 years of diabetes. At the beginning of retinopathy, minimized circulation and increased permeability of the small retinal vessels occur, which can later create little pouches in the blood vessels (or *microaneurysms*). Additionally, small bleedings and fatty "cotton wool" exudates occur later. In the more advanced stages of retinopathy, new blood vessels can develop (proliferative retinopathy), which allows blood to seep into the inner eye. Only then does a massive impairment of vision occur.

Does that mean that retinopathy does not worsen vision in the beginning?

In general, the late complications of diabetes all begin slowly, usually without any symptoms but have serious consequences. If a functional impairment such as loss of vision has already appeared, it is too late to take preventative measures. Possibly this insidious course is to blame for the decrease in vision or blindness; the latter occurs 25 times more often in people with diabetes than in individuals without diabetes, as the treatment often starts in an advanced stage of retinopathy.

Retinopathy is symptomless at the beginning: annual check ups necessary!

So anyone with diabetes — even one with good vision — needs to visit an ophthalmologist. Is that right?

That's right. You need to attend an appointment with an ophthalmologist, whose sub-specialty is retinopathy or diabetes, at least once every year. It is also advisable to undergo fluorescent angiography (in particular, after five years of diabetes), because this test makes the vessels at the back of the eye visible, hence, diagnosing the condition of blood circulation in the retina. Always ask for your explicit results and degree of retinopathy.

There might be a possibility of a specific diagnosis as follows:

(1) **No retinopathy**,

(2) **Non-proliferative retinopathy** (microaneurysms, exsudate, small hemorrhages),

(3) **Pre-proliferative retinopathy** (areas with blood supply cut off, fluorescent angiography and laser treatment necessary),

(4) **Proliferative retinopathy** (vessel's new formation, quite dangerous, risk of bleeding, laser treatment necessary).

Can retinopathy be treated?

Laser coagulation therapy can obliterate the damaged retinal vessels. With bleeding in the vitreous body (which is the most common cause of blindness in type 1 diabetes), sometimes the vitreous body may be surgically extracted. Laser therapy considerably improves the prognosis.

Has any medication been developed to prevent retinopathy?

Severe retinopathy may be slowed down by new and partially still experimental therapies or medications that have to be injected into the eye. Much more important: recently it has been proven that lowering of (even normal) blood pressure lowers the risk of retinopathy development or progression in people with diabetes. This could be achieved by ACE inhibitors (angiotensin converting enzyme inhibitors) or sartans (angiotensin receptor blockers).

Lowering of (even normal) blood pressure lowers the risk of retinopathy development

Is it correct that if retinopathy already exists, hypoglycemia can damage the eyes?

Certain studies have shown that rapid blood glucose normalization in people with diabetes who have been insufficiently treated in the past can (probably only temporarily!) worsen advanced retinopathy. This temporary harm can be prevented by normalizing blood glucose values slowly, one step at a time. There is absolutely no proof, though, that low blood glucose itself leads to the aggravation of retinopathy.

When does kidney damage take place in diabetes?

Microangiopathic damage to the kidneys first occurs after 14-20 years (in most cases) since the onset of diabetes. Constant protein excretion in the urine is characteristic of diabetic nephropathy, or kidney damage. Later, despite a long "trouble-free" period, the renal (kidney) function still can deteriorate and finally lead to kidney failure.

How can one recognize kidney problems before it is too late?

Regular urine tests for excess protein excretion (known as *microproteinuria*) are appropriate at every check-up with your diabetes physician (protein excretion values up to 15 ug/min are normal). Furthermore, blood tests are important for checking the function of the kidneys (creatinine clearance is such a test). Treating a urinary tract infection in its early stages is also important in diabetes.

How can kidney disease (associated with diabetes) be prevented or treated?

The progression of diabetic nephropathy can be delayed mainly by lowering blood pressure. However, there are some more positive steps you can take: First, if you smoke, stop smoking. It could make a difference to whether kidney disease affects you in the future or not. Second, if you have high blood pressure, see your physician immediately and tackle this problem; untreated hypertension accelerates kidney disease. Third, as just mentioned, treat a urinary tract infection soon after its onset. An excellent glycemic control and, only in the later stage, restriction of dietary protein intake may also delay the progression of nephropathy. Kidney failure used to be fatal earlier. Today dialysis ("blood filtering") and kidney transplants are life-saving therapeutic measures. *Peritoneal dialysis* (blood filtering in the peritoneum, or abdomen) seems to be better for those with diabetes than *hemodialysis* (blood filtering outside the body), which requires access to a patient's blood vessel and complicated instruments.

You mentioned that the nerves too may be damaged by diabetes. What are the symptoms of this?

Damaging the sensitive nervous system is very often in diabetes and can induce a loss of sensitivity or loss of feeling, in the feet, legs and less often in the hands. In some cases wrong or painful sensations or pain in the legs can develop. Damaging the autonomic (vegetative, intestinal) nervous system (autonomic neuropathy) can include:

- Male impotence (erectile dysfunction, ED),
- Disturbances in blood pressure control,
- Insufficient adaptation of heart frequency during physical activity, decrease of "normal" heart rate variability, high heart rate (>90 bpm),
- Disturbance of stomach emptying, diarrhea,
- Urinary disturbances; insufficient emptying of the bladder,
- Insufficient perspiration and sebum secretion in the feet.

Are there specific tests for the early recognition of diabetic neuropathy?

A simple test conducted by your diabetes physician to indicate the presence of neuropathy is to check the vibration sensitivity of your foot with a tuning fork. Your physician can also check your tendon reflexes for signs of diabetic neuropathy. A more detailed neurological diagnosis is possible using a special test (checking the forwarding speed of the nerves). After longer diabetes duration autonomic neuropathy of the heart can be assessed with heart rate variability analysis (computerized ECG analysis).

Can nerve damage be corrected or treated somehow?

The effectiveness of a few drugs to treat diabetic sensormotoric neuropathy has been proven. Some forms of neuropathy may be improved by very good blood glucose control. Early forms of diabetic autonomic neuropathy of the heart can be influenced by regular physical exercise (30 minutes at least 3-4 times a week are beneficial). Fasting has also been shown to apparently improve cardiovascular autonomic neuropathy.

How does diabetes affect the ability to have sex? Can it cause erection problems?

Diabetes can considerably raise the risk of "erectile dysfunction" (*short: ED*). Responsible for developing ED is usually a nerve damage, in most cases autonomic neuropathy. Restricted blood flow can also sometimes contribute to ED. Erectile tissue of penis has to be filled with sufficient amount of blood, which can only be done if autonomic (parasympathetic) nervous system (without conscious control) is functioning properly.

Details: The erectile tissue of the penis resembles a sponge. This sponge-like tissue is made of tubular structures that run the length of the penis and have to be filled with sufficient amount of blood that can cause a penile erection, but only when smooth muscles of erectile penis tissue is relaxed. The necessary relaxation of these muscles requires functioning nerve tissue, for neurotransmitters (kind of "messengers") to be released. Vessels' cells need a sufficient amount of messenger molecules. The production of nitric oxide plays an important role in this matter. If there is a damaged/disrupted nerve supply (autonomic neuropathy) and/or not sufficient messenger molecules are produced, the relaxation of smooth muscles can not be complete, which makes ED, i.e. erectile tissue of the penis is not filled with sufficient blood amount. What used to be taboo just isn't any more...

...and ED can be medically treated, can't it?

This is where Viagra® and its successors step in. They prevent a rapid decay of messenger molecules, which are needed in order to lead smooth muscles to relaxation. Initially was Viagra's agent (Sildenafil, Manufacturer: Pfizer) synthesized and produced for pulmonary hypertension, it dilates the vessels (of the lungs and penis), resulting in a better inflow of blood. Further developments were made by Eli Lilly (Cialis®, agent Tadalafil) and Bayer (Levitra®, Vardenafil), however the main difference is in the duration of effect. The principle is always the same: Levitra® and Viagra® act potent but brief, where Cialis® lasts longer ("weekend version").

Viagra®, Levitra® and Cialis® encourage blood flow into the penis – by slowing down a decomposition of messenger substance through an enzyme (PDE-5), which leads to vasodilatation of vessels in penis. They do not induce any sexual stimulation or sexual arousal, but they improve functionality of the penis, once aroused. However: they should not be combined with substances which cause dilatation of coronary arteries (nitrates). For this treatment you'll need a prescription. Insurance will not cover the expenses.

Are there any alternatives?

(1) **Vacuum pump:** Vacuum erection device is a cylinder with a compression ring fitted over the penis basis where it creates a vacuum that makes blood vessels dilated, leading blood flow into the penis and causing the erection. After removing the cylinder, the compression ring should stay at the basis of the penis in order to enable and secure the erection. Disadvantage: This procedure is somewhat less erotic...however, cheaper if in chronic use☺ ...

(2) **Self-injection** into corpus cavernosum: is especially useful in patients with serious nerve damage e.g. patients who went through the prostate surgery. Here is the medication injected using a needle directly into the erectile tissue

of the penis. Caution: You'll need to see a doctor first, to be given the proper dosage. Otherwise there is a danger of an overdose which could lead to priapism (long lasting, painful erection). Also for this treatment you'll need a prescription. Insurance will not cover the expenses.
(3) **Penile implant/** prosthesis.

The ED prophylaxis in diabetes would probably require good glycemic control?

The probability of autonomic and other forms of neuropathy is lower with better diabetes control.

The autonomic nervous system will also improve if you (1) exercise (at least 30 minutes, 3-4x/week) and/or (2) do modified fasting. And: Do not smoke!

Does diabetes influence woman's sexuality?

Some studies have shown that the influence in female type 1 diabetes' patients is less than in men. However in practice I recognize two common problems:

(1) Side effects of antidepression meds. Diabetes patients are more prone to depression. Standard therapy includes antidepressants, particularly selective serotonin reuptake inhibitors. Side effects: from orgasmic disorder to anorgasmia (present mostly in women). Important: if this happens, (do not exchange your partner☺) try another medication, which might not cause such serious sexuality disorders (i.e. Reboxetin/ Edronax®, Bupropion/ Wellbutrin®)
(2) Vaginal infection proneness can lead to candida/yeast i.e. vaginal thrush. Optimal diabetes control and clotrimazol local treatment helps.

Why is depression so common in diabetes?

Depression occurs often in diabetes. Highly treatable!

The probability of depression is growing proportionally to diabetes's duration and its complications. Lack of messenger molecules could be a reason, indicating some degree of nerve damage. Depression can be diagnosed if two out of three major symptoms are present – (1) grim mood, (2) indifference or cheerlessness, and (3) drive disorder or fatigue, and if another two symptoms occur e.g. attenuation in concentration, in self-confidence, appetite loss or the existence of feelings of guilt, anxiety or insomnia. Depression is highly treatable, however sometimes it may involve pharmacotherapy, if psychotherapeutic treatment should not suffice.

You stated that diabetes may affect even the large arteries. Could you tell me more about this problem?

Changes in the large arteries -- **atherosclerosis** -- are not, as you already know, specific to diabetes. However, diabetes causes particularly early and advance forms of atherosclerosis. Therefore, changes in coronary arteries of people with diabetes occur more often than in people without diabetes. Even the protection of the female hormones against coronary heart disease seems to be less effective in women with diabetes, though their counterparts without diabetes under the age of 50 are rarely affected by this disorder. In general, women who experience myocardial infarction (heart attack) before they are 50 are, usually, found to have diabetes.

Which symptoms are associated with coronary artery disease?

Angina pectoris (heart pain during physical activity, commonly known as **angina**) is typical. But doctors need to be careful in making a diagnosis. Occasionally, because of autonomic neuropathy coronary artery disease may be present in diabetes without exercise induced pain. Therefore, those who are over 40 need to have an ECG test carried out annually. If heart complaints exist already, certain special tests (cardiac stress test) are necessary in any case.

Which preventative measures are required to protect against coronary artery disease?

It is particularly important to eliminate other risk factors for atherosclerosis in particular, cigarette-smoking, high blood pressure and high blood lipids. Systematic exercise is also very important. Discuss with your doctor whether and which type of systematic endurance training is appropriate in your particular case.

What does "diabetic foot" actually mean?

The combination of neuropathy (decreased sensation), disturbances in the healing of wounds due to insufficient circulation, and finally infection can lead to the occurrence of diabetic gangrene. Amputations are 20 times more common in people with diabetes than people without diabetes. For this reason, pay special attention to foot injuries or cramps in the calves, which are typical for arterial circulation problems, as the pain sensitivity due to neuropathy is often decreased. People with neuropathy find that injuries to the foot or ill-fitting shoes do not hurt. Therefore, appropriate checks and precautions are important, as well as the proper treatment of possible injuries and pressure points and infections. Vessel closures are often not the cause of the "diabetic foot". Neuropathy is, more often, the culprit. Hence, the "diabetic foot" is rarely cold and does not lack adequate blood circulation: usually, it is hot, and has better blood supply than normal because of inappropriate widening of the vessels (nerve damages) and it is more likely to be injured. Ask your physician whether sensibility in your feet is normal. If it is impaired, be aware of a higher probability of infection after each, even a small injury and check your feet frequently!!

Are there any diseases that occur more often with diabetes?

High blood pressure and high blood lipids are almost always present in type 2 diabetes and quite often in type 1 diabetes in people with higher age. Since being vascular, they influence the prognosis, and therefore they must be treated "aggressively". If any complications are already present in these patients, following treatment goals are recommended:

Blood fats (blood lipids):

- total blood cholesterol < 175 mg/dl (< 200 mg/dl, if still no atherosclerosis)
- LDL cholesterol < 75 mg/dl (<100 mg/dl, if still no atherosclerosis)
- HDL cholesterol > 40 mg/dl (men) and > 50 mg/dl (women)
- Quotient total cholesterol / HDL < 3

Blood pressure (self-measured, targets):

- < 120/80 mmHg (< 130/80 mmHg, if still no atherosclerosis) or lower if well tolerated

And if I have type 1 diabetes, is there anything else in particular to be mentioned?

If you have an autoimmune diabetes (type 1 diabetes), there is a high probability of autoimmune hypothyroidism (disorder of thyroid gland), also called Hashimoto thyroiditis. About half of patients can develop hypothyroidism, reduced function of the gland, throughout the years. It can be associated with following symptoms: hair loss, constipation, cold intolerance, mood instability, fatigue and weight fluctuations. Thyroid produces hormone T4 (thyroxin), which is converted to another hormone, T3, by peripheral organs. These hormones are responsible for regulation of metabolism. The brain (the pituitary gland) stimulates the thyroid gland with the hormone called TSH (Thyroid Stimulation Hormone). If secretion of TSH is enhanced, in type 1 diabetes it indicates hypothyroidism, even though T3 and T4 still are in the normal range. This is where an intake of these hormones would be necessary (either pure T4 or combined T4 with T3, as pills).

Then I should do the test to determine these hormone levels?!

Exactly. At least once a year ask for the TSH test, in order to start the treatment on time.

Could you summarize the routine tests necessary to assess regular late complications of diabetes?

Blood analysis (HbA_{1c} should be done *3-4 times annually*, kidney function parameters, blood lipids);

At least *once a year*, the following tests need to be performed:

1) Retina test with an ophthalmoscope (after the tenth year of diabetes, possibly also fluorescent angiography);
2) Urine analysis for protein (microproteinuria);
3) Blood test for TSH, check up for the functions of the thyroid gland

Test of foot pulse; checkup of your feet; Test of sensitivity to the feet (tuning fork, tendon reflexes).

Much more often, usually with *each ambulatory visit:*

- Weight
- Waist circumference

("normal values")	*in Europe*	*in the US*
Women	*<80 cm*	*<88 cm*
Men	*<94 cm*	*<100 cm*

- Blood pressure needs to be checked much more often than with each ambulatory visit! Even a slight increase in blood pressure can be a sign of kidney damage and needs to be treated properly and requires blood pressure daily self-monitoring. An increase in blood pressure is present at least in 80% of all diabetes patients.
- When increased microproteinuria is diagnosed (over 20 ug/min, particularly from collected urine) or retinopathy becomes advanced, an increase in blood pressure is virtually found in each case. Self-adjustment of blood pressure through medical treatment can be learned in a special *hypertension training education model*, as well as self-monitoring of the blood pressure (time needed: 9 hours of training, spread over 3 evenings). Normalization of blood pressure often slows down the advancing of retinopathy and kidney damage. Normalization of your blood glucose and blood pressure values is the best precaution to prevent or delay late complications.

Self-monitoring of blood pressure is an absolute must in any kidney damage from diabetes (increase in microproteinuria). If you think you have "normal" blood pressure despite increased protein secretion, we would strongly suggest that you measure your blood pressure for 24 hours with "ambulatory 24hrs blood pressure monitoring (ABPM)" system. Using this method, it can often be proven that blood pressure values are increased during the night or that the physiological and desired nightly blood pressure decrease ("night dipping") no longer exists; that too can be influenced by adequate medication.

Hypertension training Emphasis: diabetes mellitus and secondary hypertension
Teaching unit I
– Causes of high blood pressure (hypertension) – Causes of hypertension in diabetes mellitus – Detecting kidney damage in diabetes mellitus – Non-drug treatment of hypertension, lowering of natrium intake and of body weight, RESPeRATE® – Drug therapy for high blood pressure (outline) – Blood pressure self-monitoring, discussion of results, treatment self-adjustment – 24-hour ambulatory blood pressure monitoring (ABPM)
Teaching unit II
– Summary: causes of high blood pressure – Significance of secondary hypertension in diabetes – Most common mistakes in self-monitoring of blood pressure – Hypertension drugs and their characteristics: - ACE inhibitors - Sartans - Renin inhibitors - Calcium antagonists - Diuretics - Beta-blockers - Vasodilators - Centrally acting substances – Discussion of blood pressure values self-measured by the patient – Therapy adjustment
Teaching Unit III
– Summary of drug treatment for hypertension – Log book discussion and therapy adjustment – Lowering of blood pressure as prevention of late complications in diabetes – Stopping antihypertensive therapy: when and for whom is it possible and beneficial? – Therapy adjustment in a hypertensive crisis – High blood pressure and other risk factors of coronary artery disease – Discussion

Fig 14.1. Contents of Ambulatory Hypertension Training for Diabetic Patients with Late Complications

15. FIT and Pregnancy with Type 1 Diabetes

Will my children have diabetes?

Contrary to earlier assumptions the probability that children of people with type 1 diabetes will also have diabetes is low. In women with type 1 diabetes the probability that their children will have diabetes is 1-3%; for men who have type 1 diabetes the probability is slightly higher (about 6%). On the other hand, if you have type 2 diabetes, the probability of inheritance is much higher, about 50%.

What risks are associated with pregnancy in women with type 1 diabetes?

The main risks are: possible congenital deformities, malformations and a high birth weight, 9 pounds or more ("macrosomia"). Increased blood glucose in the mother during the first 3 months of pregnancy leads to the development of congenital malformations, mostly in the nervous and circulatory systems. The probability of this occurring in general population is about 1%. In mothers with diabetes who are insufficiently treated the percentage increases to approximately 7%.

And what if excellent glycemic control is achieved during pregnancy?

It has been proven that near normal glycemic control from conception onward, especially in the first trimester (at least during first 9 weeks, during which the development of internal organs (i.e. "organogenesis") takes place) brings the malformation rate of children of diabetes mothers closer to that of children of non-diabetes mothers (1%). Apart from near-normal glycemic control I would advise you to take folic acid (protects from malformation of central nerves system) and antioxidants: tocopherol (vitamin E, at least 2000 mg/day) and ascorbic acid (vitamin C, at least 1000 mg/day). All of these substances are innocuous, and are proven to prevent the malformation of the fetus, however only in animal studies up to now.

Malformations develop only during the first trimester of pregnancy. Why is it very necessary to achieve near-normal glycemic control during the whole pregnancy?

Poor glycemic control during the last trimester is associated with an increased risk of premature death of the fetus. In the last trimester when the fetus begins to produce its own insulin, the fetus produces too much insulin. The baby "reacts", so to speak, to the increased blood glucose of the mother and "wants" to normalize it. As a result of the increased glucose transferred from the mother and the subsequent increase in insulin production by the fetus, "glucose-insulin fattening" by the fetus and the birth of a large infant (macrosomia) result. By the same mechanism, hypoglycemia often occurs after birth in infants of mothers with diabetes. If the fetus produced a lot of insulin to compensate for the mother's high blood glucose, after birth the infant will become hypoglycemic because his/her insulin level is too high after leaving the mother's womb.

What are the risks for women with diabetes during pregnancy?

The risks for the mother with diabetes are related to her late complications of diabetes, i.e. changes in the blood vessels and impaired kidney functions. It is particularly difficult when there is high blood pressure is pre-existing.

What consequences occur in practice?

Normal glycemic control during the entire pregnancy is hardly attainable without profound patient education. Not only a FIT-training is necessary; additionally you will need a specific *pregnancy- and delivery education module* to attend before your 28th pregnancy week. The following health professionals need to be involved in the care of a pregnant woman with diabetes: obstetrician (approximately 10 visits during the entire pregnancy), an ophthalmologist (at least two visits during the entire pregnancy) and a diabetes specialist who not only trains the patient for self-treatment but also supervises the therapy. A pediatrician needs to be available to take care of the newborn. Optimal care is usually available from specialists at special training centers. It is important to plan your pregnancy! The best contraception for women with diabetes are barrier methods (condoms and diaphragms) which are best in combination, or new low-dosage contraceptive pills (which are, however, probably not good for smokers, or those with vessel damage). Discuss your contraception needs with your physician.

Is an in-hospital admission necessary during pregnancy?

That depends on your glycemic control and how you carry out your therapy. If you are FIT trained, you have already reached near normal glycemic control before pregnancy (normal HbA_{1c}), and you are well trained in the choice of insulin dosage specifically during pregnancy, you will be admitted to the hospital only after contractions start for delivery if no pregnancy complications occur. Once more, additionally to FIT training we recommend specific *pregnancy in diabetes and delivery module* (one afternoon) where your partner is also welcome. We recommend that pregnant women attend our clinic weekly during organogenesis and in the last trimester (time needed: 10 minutes per week plus travel). These short visits allow avoiding unnecessary in-patient admission. Caesarean section, often required before, can be avoided when there are no problems with the child or the birth process.

How often does a pregnant woman with diabetes need to visit the obstetrician?

Approximately every three weeks. Even more frequently (every two weeks) after the 30th week of pregnancy, to exclude macrosomia. Ultrasound tests of the fetus will be performed on these visits. Its development will be observed and it will be checked for malformations. A specific test via ultrasound to recognize malformations of the baby need to be carried out by an obstetric center with experience in pregnancy with diabetes between the 20th, and at the latest, the 23rd week of pregnancy via ultrasound.

Which strategies of insulin treatment are particularly good during pregnancy?

FIT, naturally! An independent choice of the insulin dosage according to the changing insulin requirement during pregnancy is an absolute prerequisite for an appropriate insulin treatment. During pregnancy, FIT is strongly recommended. Immediate corrections of each hyperglycemia are indispensible. You need to talk to your diabetes physician, informing them immediately about your pregnancy plans to establish new therapeutic targets early enough.

Why is that so important? And if I can manage FIT, my treatment will not change?

Avoid unplanned pregnancy

Nothing major will change. In any case, you need now to aim for real normoglycemia. For best outcomes, you need to:

1. **Avoid unplanned pregnancy.** Even prior to conception start taking folic acid as well as antioxidants as described.
2. **Calculate your MBG daily,** in order to assess daily your glycemic control. Usual good diabetes control (MBG 110 – 160) is not sufficient for pregnancy (MBG < 100 (110) and below 90 (100) mg/dl in the last trimester)

3. **Lower your blood glucose target points for correction.** Usually, target levels of 90 mg/dl before meals and up to 120 mg/dl after eating are set. In pregnancy, blood glucose values of 65-90 mg/dl will be tolerated without an "upward" correction. This corresponds to an achievable aim of daily mean blood glucose under 100 mg/dl. If you have had repeated episodes of serious hypoglycemia in the past, a somewhat higher target area can probably be chosen. However, mean daily blood glucose over 120 mg/dl is not acceptable. The target area for MBG in the last trimester is more often established at less than 90 mg/dl. These blood glucose values correspond to low-normal values of glycated hemoglobin (HbA_{1c}). With "normal glycemic" control you are trying to prevent congenital malformations in the first trimester and the development of a large infant, or macrosomia in the last trimester and its associated complications. Your personal involvement and FIT training are decisively meaningful for the fate of your baby especially at the beginning and end of your pregnancy.

4. **Double your efforts for self-monitoring.** Instead of 4-5 measurements of blood glucose tests daily carry out 8-10 (i.e. almost "hourly") measurements. Fewer than 6-8 measurements are, in our experience, not enough. Measure your blood glucose much more often postprandially, 1-2 hours after meals because that is when the greatest danger of unacceptably high values exists. In particular, when you try to eat less to avoid an excessive weight increase measure daily ketones in your urine. During pregnancy, it is not advisable to begin a weight loss diet. A weight increase of approximately 25-30 pounds during pregnancy is considered normal.

5. **Visit your diabetes physician more often for a check-up.** During the first and last trimesters, a weekly appointment is recommended. Even if your glycemic state is very good, frequent check-ups are needed because of the constant insulin requirement changes during pregnancy. Especially if late vascular complications already exist, your blood pressure values, protein excretion in the urine, kidney function, and ocular changes need to be monitored frequently.

6. **Be prepared for hypoglycemia** with such a low blood glucose target level. Contrary to earlier fears, hypoglycemic episodes in the mother do not harm the child. Nonetheless, do not endanger yourself. Avoid serious hypoglycemia by testing frequently.

How does the insulin requirement change during pregnancy?

During the first three months of pregnancy the daily insulin requirement falls slightly, followed by increase from the 14th to 17th week because of the increased production of placental hormones. The highest daily insulin requirement (two to three times the values before pregnancy) exists toward the end of the pregnancy. In the last few weeks of pregnancy, the insulin requirement can decrease indicating that the placenta function and pregnancy hormone production is decreasing. After giving birth, the insulin requirement falls drastically to about 30-50% of the value before birth and remains somewhat lower than before pregnancy. And usually remains slightly lower during the whole breast-feeding period.

All of these factors need to be taken into consideration when adjusting insulin algorithms. During pregnancy, prandial and basal insulin dosages are increased step-by-step according to the daily mean blood glucose (MBG) values, MBG should remain under 100(110) mg/dl and under 90 mg/dl after the 29th pregnancy week as discussed. Mean blood glucose in healthy women in pregnancy is around 85 mg/dl. Only if you have a history of serious hypoglycemia, you are allowed to keep your MBG 10-20 mg/dl higher, provided no signs of macrosomia are shown by ultrasound.

What is the procedure during birth?

The insulin requirement is halved or reduced even further during and after birth. The algorithms for basal and prandial insulin need to be reduced to approximately 30% of the usual value during late pregnancy.

Do the physicians and other health professionals care for the woman during delivery?

Yes, but it is advantageous if the woman can monitor and control her own blood glucose. Of course, like all expectant mothers, you would need to be prepared during late pregnancy. Prepare everything on time, for delivery. "Extended minimum FIT equipment" needs to be ready. When labor begins, reduce the basal and prandial algorithms that you have used up to now to one third, take blood glucose tests at approximately one-hour intervals and try to remain close to the normo-glycemic level. Please discuss details with your FIT physician. If you are exhausted and cannot look after yourself communicate this clearly to your physician and your midwife. Assume the responsibility for yourself as soon as possible after the birth of your baby when you are feeling better again. You will choose your insulin dosages again according to new reduced algorithms.

What are the guidelines for using an insulin pump during pregnancy?

The insulin dosage for a pump is virtually the same as for injections. During and after birth, the same principles need to be followed. The algorithms used until start of contractions need to be reduced by half or even to approximately one third.

Rose 39th pregn. week

INTERNATIONAL RESEARCH GROUP ON FUNCTIONAL INSULIN TREATMENT
Medical University Vienna
kinga.howorka@meduniwien.ac.at
www.diabetesFIT.org

39th week

Name :..
Birth date:....................Phone:...........
Address:..
E-mail...
Diabetes.since:...............Wt.:.............
FIT since:........... with O injections O pump

I BASAL (fasting insulin): AM **15 NPH** / **5 NovoLog** U N **Midday 15 NPH** PM **16 NPH** U S PRANDIAL: for 1 carb choice (15g) = **5 NovoLog** U U	**Target for correction of aberrant BG values:** Fasting/pre-meal: 100 mg/dl (or **85**) After meals: 1h<180 (or **130**), 2h< 140 mg/dl **Target range for MBG: from .... to 90..... mg/dl**

L CORRECTION: 1 U rapid ins. lowers my BG by approx.- **10**; 1 carb raises my BG by approx.+ ....mg/dl
I EXAMPLE: carb choices/CHO: ....
N insulin (U): ....

	TIME	1	2	3	4	5	6	7	8	9	10	11	12	1	2	3	4	5	6	7	8	9	10	11	12	Total Daily Dose	
							AM													PM							
MON 26 Apr	Basal	Basal NPH							15						15									16		46	108
	Bolus	NovoLog							5		15				16			10					16			62	
	BG								72		78			90		82		80		91		101		86		MBG	85
	Carbs/CHO										3				3			2					3			11	
	Comments	Ketons neg.																									
TUE	Basal								15							15								16		46	93
	Bolus								15						15						15			2		47	
	BG		65						81		87				72		105		68		79			119		MBG	86
	Carbs/CHO										2				3					1,5	3					10	
	Comments	Ketons neg.																									
WED	Basal							15						15									16			46	106
	Bolus							5			15			20		3				15			2			60	
	BG		77						82			88		75			127		65	62				102		MBG	84
	Carbs/CHO										3			4						3						10	
	Comments	Ketons neg.																									
	TIME	1	2	3	4	5	6	7	8	9	10	11	12	1	2	3	4	5	6	7	8	9	10	11	12	Total Daily Dose	

Fig. 15.1a: Rose expects delivery within short time.
How do you feel about her glycemic control? Which specific measures has she taken for assuring the best outcomes? Are they appropriate for every pregnant woman with diabetes?

Fig 15.1.b:

This patient is obviously quite motivated...

Pregnancy specific is:

- Higher number of blood glucose self-measurements (at least 8/day),
- Measurements during night,
- Lowering of blood glucose pre– and postprandial target points for corrections,
- Lowering blood glucose target for MBG (Rose's MBG is less than 90 mg/dl!)
- Perceive, note and shorten all the time intervals where blood sugar is over 85 mg/dl (potential stimulation of baby's insulin production)
- Insulin dosage adjustment: doubling of total daily dose requires an increase of algorithms for basal and prandial insulin, numerical decrease of correction algorithm:
  "1U lowers my blood glucose by -10mg/dl"
- Labor induction can be achieved at term by the reduction of food intake or deliberate fasting; Rose is almost there...

Such pronounced decrease of MBG cannot be always attained without problems. In hypo-risk patients a slight deliberate hyperglycemia has than to be accepted. But, the danger of severe hypoglycemia is in the last trimester quite low -- due to an increasing insulin resistance. Therefore, slight overdosing will usually be well tolerated and without serious hypoglycemia.

Insulin analog NovoLog® was tested and shown to be appropriate for the usage during pregnancy. In general, the use of rapid acting insulin analogs is associated with less risk of hypoglycemia.

And what if a caesarean section is needed?

Follow the same principles. If you need an epidural anesthesia, inform the anesthesiologist that you have already reduced your insulin algorithms and by how much. Show him/her your log book.

Diabetes during birth sounded very complicated up to now. Is a glucose or insulin i.v. still necessary?

If you follow the FIT principles the answer to this question is "no". A glucose infusion consisting of approximately 2-3 carbs altogether, approximately 20-40 grams of glucose is only necessary during birth when your high basal rate is still in effect requiring glucose therapy. This infusion would be unnecessary if you know that your basal rate is too high and you consume correctional 1-3 carbs as glucose tablets.

It is logical to stabilize this blood glucose condition with glucose, but women cannot take food during labor – is this correct?

No one should eat before "surgery". This is to prevent vomiting. In principle, consuming a few grams of glucose will not hurt you. Honey (has been given to the pregnant women during labor for centuries) or glucose is absorbed so quickly that vomiting hardly ever occurs. Your physician will not oppose taking a few glucose tablets in this situation. Ask him/her for advice.

And after delivery? What is necessary to monitor my child?

Even if your child is healthy immediately after delivery, the danger of hypoglycemia exists in the first few hours of the baby's life. The only means of preventing this is to closely monitor the child's blood glucose at least every few hours. If the blood glucose values are too low (hypoglycemia in the newborn is usually defined as blood glucose value below 30 mg/dl), they need to be corrected. For the sake of safety this procedure needs to be done at the hospital where the delivery has occurred. Early breast feeding is strongly recommended. The risk of hypoglycemia in the infant stops approximately 24 hours after the birth. The child needs to be checked by an experienced pediatrician for malformations or complications associated with diabetes.

Now let's summarize the problems of pregnancy in diabetes:

1. **Plan**: Pregnancy in type 1 or pregestational diabetes needs to be planned in advance so that optimal glycemic control can be achieved already before conception.
2. **Normoglycemia**: Throughout pregnancy, normoglycemic control is necessary to prevent congenital malformations and associated complications in the newborn, promote normal development of the fetus and prevent a high birth weight/ macrosomia. Mean daily blood glucose should stay below 100 (110 or 120) mg/dl, and even below 90 (100) mg/dl, after the 29th week of pregnancy.
3. **Self-monitoring**: The effort devoted to self-testing needs to be doubled. At least 8-10 daily blood glucose self-checks are recommended in type 1 diabetes. Currently, multiple injections or an insulin pump are considered absolutely necessary during pregnancy.
4. **Experts**: Interdisciplinary supervision of the mother with diabetes is necessary at an experienced diabetes clinic involving the cooperation of the diabetologist, obstetrician, ophthalmologist, and pediatrician.
5. **Investigations**: Essential is the malformation screening (in the second trimester, pregnancy week 20-23), as well as quite often ultrasound checks after the 30th week of pregnancy (every two weeks) in order for development of the fetus to be observed (macrosomia?).
6. **Checkups**: Weekly check-ups by a diabetes specialist experienced in supervising pregnancies are recommended, at least during organogenesis and the last pregnancy trimester. Incipient signs of macrosomia (ultrasound) have to be counteracted with lowering of the target for MBG and hyperglycemia corrections.
7. **Blood pressure**: Particular supervisory precautions are important when late complications of diabetes are already present in the mother. Tight blood pressure control is necessary.
8. **FIT algorithms**: Secondary adjustments of the insulin dosage throughout the pregnancy as well as during the delivery by the mother herself are desirable and necessary. During pregnancy, the daily insulin requirement increases slowly but decreases drastically during and after birth.

Our patients have shown that FIT program combined with specific pregnancy module provides the same outcomes as in "normal" non-diabetic pregnancies. Macrosomia or postnatal hypoglycemia is rare (in less than 10% of our childbirths). Our results are the best ever published results in the world regarding diabetes and pregnancy outcomes (for our publication see Diabetic Medicine, 2001).

A final tip: Breast-feeding is recommended to all women. Recently, it was proven that exclusive and relatively long breast-feeding (6 months and longer) partially protect children from a later manifestation of type 1 diabetes. Further studies on breast-feeding and type 1 diabetes are ongoing and important. Women who have type 2 diabetes, should also breast feed their newborns, which protects them from gaining weight in adult life and thus protect them against type 2 diabetes as well. Insulin needs can decrease in lactating diabetic women.

16. Functional Insulin Treatment in Children and Adolescents

Is FIT a good idea for children and teenagers?

Of course, FIT is not only suited for grownups but for children as well. However, children need to be under the care of at least one parent or a pediatric endocrinologist who understands special needs of diabetic children during their growth and puberty stages. After the age of 10, children can attend FIT (with special modifications) where they would be taught how to maintain treatment on their own. Several "empowered" American parents of children ages 7 and older have trained their children on FIT principles in our program with great results.

What are those great results?

Strong "catch up" growth in the first few months after FIT training. Children, whose HbA1c's are not near the normal range, have stunted linear growth. Some may even be overweight because of the need to eat carbs to chase their insulin dosages. The children have regained good scholastic averages because they feel better and are able to stay in mainstream education at school.

Children should get their own FIT training at the beginning of puberty and another refresher shortly before becoming an adult.

So these children and teenagers also have near normal blood glucose?

At least it should be even easier for the parents to balance blood glucose levels with FIT. If the parents can see the trend over one or two days and accurately adjust insulin for these "growth spurts" then the child is living in proactive instead of reactive care and can benefit from safe, tight control and the freedom to eat as much of the childhood favorites that they want. Most FIT youngsters are within the 1.5% to 2% of the upper reference range for normal HbA1c, without strict diets, insulin schedules or serious hypos.

I've heard many children with diabetes have to repeat a grade in school, why?

Conventional insulin treatment is often insufficient. If you are using a delayed acting insulin to cover food, this is physiologically incorrect. Many children have high post meal values after lunch because they are prescribed NPH to cover their lunch – which is injected as the morning basal dosage. High blood glucose has side effects, like frequent thirst and urination, blurry vision, sensitive attitudes, lethargy. All these play a role in the child's ability to succeed at school and socially.

So they feel tired and thirsty after eating lunch?

Yes, they report that they feel tired and need to drink excessively to recover from dehydration caused by the kidney's flushing the high post meal blood glucose into the urine. Parents report improved learning abilities in children primarily because the kids feel better when their post meal blood glucose are 160 mg/dl or below.

How do schools feel about all the extra work involved in caring for children who are partially trained in FIT?

In my experience, the schools are very enthusiastic about their new found "power" to keep their students in the mode of "learning" and simultaneously feeling good with fewer days missed due to poorly controlled diabetes. In fact, many of the schools of our FIT kids allow the children to test their blood in the classroom. This keeps the child's focus on learning and helps them "normalize" their lives.

What about very young children, what happens at lunch in school?

Of course, a health aide or school nurse has to participate on the team with the student, teacher and parent. Several systems have been set up to help the child stay within mainstream education at school. The schools report great satisfaction in “not” having a preset “schedule" to feed the child in order to keep them in a safe place; and great relief that they no longer need to police the child's food consumption and "make them eat". Generally, there is less pressure. They need to have a system where a "buddy" or lunch monitor observes which foods are eaten so that the appropriate insulin dosage is administered after lunch or even after the lunch recess. The American Association of Diabetes Educators and a very compelling website for children, www.childrenwithdiabetes.com have great resources and guides for children with diabetes in school settings. You need to know your rights and work as part of the school's team.

That brings up another concern: what about gym or physical education?

Again, it takes a team approach and focus on what is best for the child. Children with diabetes need to test their blood glucose before and after strenuous exercise to avoid lows. Many parents prefer to give their children a "sports drink" with electrolytes to sip during exercise containing 1.5 to 2 carbs, based on their blood glucose reading. This gives the child the energy they need without a huge increase in blood glucose during exercise.

What if the child has a high blood glucose of 250 mg/dl or more before exercise?

In this situation, the child needs to check for ketones in urine, as always. If ketones are present, the child needs to refrain from exercise and be given plenty of sugar-free fluids or water and a usual and necessary insulin bolus to correct their blood glucose to their recommended target. They need to refrain from exercise until ketones disappear. If no ketones are present, the child can exercise as usual after a weak correction or without eating carbs before or during sports starting with slight elevated blood glucose.

Is it true that children are somewhat immune to the effects of high blood glucose as far as probability of late complications is concerned?

This has not been proven yet. Children have been diagnosed with early stages of kidney damage and retinopathy. For the sake of the child, blood glucose targets are set a little bit higher say 120-145 mg/dl in children 7 years old or more to protect their bodies and at the same time avoid serious hypoglycemia.

Is it true that severe hypoglycemia causes brain damage?

There is no sufficient evidence on the short–and long-term effects of severe hypoglycemia with unconsciousness or seizures. In any event, it is best to avoid the reaction if possible and treat low blood glucose in children below 100 mg/dl immediately!

Is treatment different for a child?

Not really. Glucagon is the treatment of choice for severe hypoglycemia when a child is unconscious, although a half dosage is generally enough in smaller children. FIT children/teenagers have their own algorithms for how much 1carb choice raises their blood glucose and how much 1 unit or even a half unit of insulin lowers their blood glucose.

Don't children mind all those injections and finger pricks?

We give children the ability to live freely, with more shots or a pump, or to follow a strict dietary regimen and inject twice daily only. No child has ever wanted to follow the regimented diet. Kids like being "normal" too. In fact, several of the children comment, "the shots don't hurt but the headaches and tummy aches do". Pumps are becoming now recommended for children even under the age of 10. However, If it malfunctions or the catheter becomes blocked they are unable to react appropriately. So only in a continually supervised life, the younger child can succeed with a pump. A child should never be "forced" to wear a pump if they do not wish to.

Why do they have headaches and tummy aches?

In the past, primarily because of the temporary high post-meal blood glucose. Now, FIT educated parents no longer feel "guilty" when their children are occasionally high because they now understand how to treat the situation safely and effectively by injecting the right amount of insulin, in the right place, at the right time and for the right reasons. Stress, excitement, and growth hormone all cause blood glucose to increase, at no fault of the caregiver. That is life! It is empowering to be able to correct these blips outside of the "target" area for the child both accurately and safely.

So, the children and their parents both like FIT?

In our experience, it has been unanimous. Every parent and child has the right to choose the insulin therapy of their choice, rediscover their real appetites, sweet tooth, and overall "spontaneity" in their lives. Why give up Holiday treats, celebrations, or exercise when you know what to do to enable you to enjoy those "happy" moments?

So I imagine that the schools like it when children with diabetes can eat cupcakes for birthday celebrations and other candy or treats without restrictions?

Yes, this is true. However, the child is the one who is likely to take most benefit, both physically and psychologically by participating in all that life offers, safely. We believe that real FIT education program with adequate "insulin games" is necessary for each kid, to recognize and accept limits for safe insulin replacement and for learning to deal with blood glucose blips or hypos. Once, at the beginning of puberty and another refresher teaching FIT course before becoming a "real" adult.

Never ever skip basal insulin injections!

What about sick days?

It is not very different from the guidance for adults, although, some children may go low instead of high when they are ill. It is easier for the children and for the parents when the child does not have to eat to chase insulin if they have the flu, diarrhea and/or are vomiting. Managing illness becomes more like it is for any other child without diabetes. The parents still worry, but act accordingly based on their knowledge and the "feedback" from the child's FIT log. In all situations, children need to be protected from serious hypoglycemia. And: Never ever skip basal insulin injections!

How do you do that if the child is vomiting?

This is the problem with conventional therapy: children "must" eat even though they are ill. This is not necessary with FIT. It's just as effective without the nausea. Fluids are more important, of course.
And, once more: Never ever skip basal insulin injections!

In your experience, at what age is a child able to take control of their FIT regimen?

The child needs to be able to do simple math and be willing to take the responsibility. Most are willing to do so, as they like the "ownership" of the treatment and the ability to eat without asking Mum or Dad. The earliest we would recommend converting children to self-treatment would be at the age of 10 years, depending on their intellectual capacity and desire to do so.

At latest, they should get their FIT training at the beginning of puberty and another refresher FIT course should take place shortly before becoming an adult.

What tests do you recommend to keep children in safe territory and to avoid complications?

The same tests as adults, meaning annual screening for retinopathy, quarterly HbA1c, microalbuminuria, etc.

It's a lot to know!

Yes. You need to know your options and preventative steps and solutions, or risk the consequences. Daily self-monitoring is a small price when compared to the consequences of uncontrolled diabetes: blindness, amputations, kidney and heart failure and premature death. Of course we would all like to see a cure for prevention of diabetes, but until then, this is still the best available way.

Do you recommend telling children about the "consequences"?

I think this depends on the age and maturity of the child. I prefer to focus on the rewards of good choices and not the consequences of bad care. Many people have something they need to treat (e.g. myopia needing contact lenses, allergies requiring epinephrine, asthma requiring inhalers). This child has diabetes but is still a child. Children with diabetes should have the freedom to run, play, eat candies, grow, learn, feel good and sleep without fear. We try to maintain good healthy lifestyle models while listening to the child's concerns and helping them "cope."

Do you recommend support groups for children/adolescents?

It really depends on what the child needs and desires. There are good support groups and poor ones. If the child does not wish to go, don't force him/her. Often, the parent may need the support more than the child. Children are very flexible. Children certainly need to know what's going on in their bodies and be able to tell those around them about their needs. They often need to have strong voices as many "adults" do not understand children's needs. One of the advantages of going to a special pediatric endocrinology center is the contact with other children with diabetes and their families in community support groups.

Are there any "recipes" for puberty in type 1 diabetes?

I have really a big respect for parents of puberting kids with diabetes. Social group pressure, depression and emotional jumble and confusion make the treatment and daily routine sometimes very difficult.

Following facts are important and should be communicated:

(1) Puberty is difficult for everybody and for those with diabetes even more. But, finally, one day you get through.

(2) Glycemic control during puberty is very important. You can usually trace in those with late complications already in their twenties how bad the diabetes control during their puberty really was. For prognosis therefore is the communication with your therapist and focus on every day solutions relevant.

(3) Sometimes, due to hormonal situation very high dosages are necessary.

(4) FIT allows flexibility and thus potentially adaptation to the eating habits of other same-aged people. But, only if FIT premises are fulfilled: self-monitoring, frequent correctional boluses, and meal-related insulin replacement. Once more: Only if those FIT premises are fulfilled.

(5) Even already "intensively treated" children should get their own FIT training at the beginning of puberty – at the very latest – and another refresher FIT course shortly before becoming an adult.

17. FIT and Type 2 Diabetes

Is Functional Insulin Treatment equally suitable for people with type 2 diabetes using insulin?

Yes. People with type 2 diabetes can use FIT successfully if they agree to inject insulin with every meal and to correct blood glucose if necessary.

Are there any serious differences between treatment of type 2 and type 1 diabetes?

Type 2 diabetes is always associated with increased blood lipids and blood pressure

Main treatment differences are based on the following differences in mechanisms of the disease: (1) in type 1 diabetes there is a lack of insulin, (2) in type 2 diabetes there is sometimes "enough" insulin but this insulin does not work sufficiently. It is not the lack of insulin, but the insulin resistance, that is most relevant. The deteriorated insulin action is compensated for by the body with an increase of insulin production. Insulin resistance, therefore, often results in a number of disturbances grouped together as *"the metabolic syndrome"*. Often, in type 2 diabetes we find the following mechanisms or components of **metabolic syndrome**, also called "**the deadly quartet**":

- central (abdominal) obesity, decrease of insulin action, increase of insulin resistance and of insulin levels,
- increase of blood lipids or cholesterol,
- raised blood pressure,
- eventually: impaired glucose tolerance and type 2 diabetes mellitus.

Metabolic syndrome is based on genetic predisposition. Its clinical manifestation is provoked by low physical exercise and high calorie intake.

Does this mean that people with type 2 diabetes may have additional diseases other than "only" diabetes?

Metabolic syndrome – "the deadly quartet"-- kills via atherosclerosis

Yes, this is virtually always the case. Therefore, the other components of the metabolic syndrome need to be appropriately treated. Not only does the blood glucose need to be brought within target range but, also, blood pressure and lipid metabolism need to be optimized in order to slow down the development of atherosclerosis. Atherosclerosis is a closing of vessels or arteries. It is a primary cause of death (up 70%) in the majority of western populations. It is more severe in type 2 diabetes and metabolic syndrome. I consider educating people with type 2 diabetes about their treatment options to be critically important! And, I recommend the following group education modules for them:

- Education on blood fats (*Hyperlipidemia Group Module*),

- Education on blood pressure (*Hypertension / Kidney Disease Prevention Group Module*),
- Weight management education: (the "*Slim'n'Fit teaching course*" and/or "*Slim-dates*", specific module comprehensive for people with metabolic syndrome).

Details on these education modules you will find at the end of the chapter on late complications.

Isn't there any simple solution to all these problems?

Yes there is. Change your lifestyle! In other words, eat less and exercise more often.

Are there any differences in insulin treatment and glycemic control?

In people with type 2 diabetes, sufficient insulin production often facilitates the treatment. In combination with insulin resistance, this has some advantages compared to type 1 diabetes: (1) less glycemic instability (less dramatic blood glucose fluctuations), (2) less risk of severe hypoglycemia with unconsciousness. These facts make FIT in people with type 2 diabetes easier to manage. The main difference applies for the replacement of the basal insulin requirement.

Basal replacement? The function of basal insulin is to ensure stability of blood glucose between meals and good fasting blood glucose values. Is it necessary to use insulin to cover the "basis" in type 2 diabetes?

For someone with type 2 diabetes, basal delayed-acting insulin is not required provided they have acceptable fasting blood glucose values. Generally, basal insulin requirement with FIT in type 2 diabetes can be covered as follows:

1. **Not at all**. The treatment can be based solely on diet and exercise with systematic use of rapid-acting insulin for meals and blood glucose corrections; no tablets/pills, no delayed-acting insulin. This is only acceptable as long as fasting blood glucose values do not exceed 130 – 140mg/dl.

2. **With oral agents**, i.e. with tablets. Rapid-acting insulin is used for meals and corrections as usual. Different types of medication have different actions:

2a.) Metformin (belongs to **biguanides)** is particularly recommended to cover the basal requirement. The action of these tablets is mainly through decrease of glucose production in the liver. Metformin is successfully used particularly by people with type 2 diabetes who are overweight. It has an additional, favorable, effect on appetite control. The UKPDS, United Kingdom Prospective Diabetes Study (1998) is the most important, relatively recent large-scale study of type 2 diabetes. It demonstrated that metformin improve the long-term outcomes of type 2 diabetes, delaying complications by 32% and improving survival (by 36-42%). Metformin can only be used as long as kidney function is not impaired (serum creatinine under 1.5 mg/dl), otherwise there is an increased risk of lactacidose, which is a dangerous albeit seldom complication.

2b.) Sulfonylureas. Long-acting sulfonylureas like Glimepirid (Amaryl®) increase insulin production in the pancreas. It is possible to combine sulfonylureas with insulin, either for "basis" or with rapid-acting insulin for meals. However, in contrast to biguanides, sulfonylureas can cause hypoglycemia that can be considerably worse than those caused by inappropriate insulin use. Furthermore, using sulfonylureas for basal insulin requirements makes the use of rapid-acting insulin to correct hyperglycemia difficult because blood glucose drops "spontaneously" with sulfonylureas anyhow. In contrast, rapid insulin releasers stimulate the beta cells only over short period of time and only if blood glucose is high. It is useful for prandial use.

2c.) Insulin sensitizers. This new and promising pharmacological group is reported to improve the insulin action. It has been reported, that Pioglitazone has more advantageous impact on blood lipids than Rosiglitazone.

2d.) Incretins and their analogs (gliptins) are intestinal hormones, which increase insulin and decrease glucagon production after a meal. They enhance satiety after a meal lowers blood glucose. With less food, they allow much better weight control. (Januvia® / Sitagliptin/ MSD, Galvus® / Vildagliptin/ Novartis, Byetta® /Exenatide / Eli Lilly).

3. **Delayed-acting insulin**. Like in type 1 diabetes, in type 2 it is also possible to cover the basal needs using delayed acting insulin (NPH, Levemir® or Lantus®). A very long-acting insulin (Lantus®) in the morning and an intermediate insulin late in the evening (NPH / Levemir®) allows a person to lower her/his requirements for delayed acting insulin. Such combined "basal" improves fasting values. The basal substitution can also be carried out by combining insulin with biguanides or sulfonylureas, as reported. This enables the insulin dosage to be kept much lower.

On the other hand, could you use some residual capability to own insulin production for meals and replace only the basal?

Yes, you can. And it makes the treatment quite easy – a good option for insulin initiation. So you use just one injection of long acting insulin. NPH, Levemir®, or Lantus® have all been used for this purpose. Lantus®, with the longest action, seems to be very useful for such therapy regimen. Later on, as insulin need increases and the patient´s own production decreases with time, meal related and correctional boluses should be integrated into the treatment, as well. After all, in type 2 diabetes you will finally find the full FIT option to be necessary. Eventually, separate replacement of prandial, basal and correctional insulin will become indispensable.

Are there any peculiarities of insulin use for meals or blood glucose corrections with FIT and type2 diabetes?

There are some indications that, in type 2 diabetes, the "early" response of insulin production on the meal is deteriorated. That is, whereas in a healthy person a high insulin level is reached within a short time after carbohydrate ingestion, in people with type 2 diabetes, this process is slowed down. The new rapid-acting insulin analogs, e.g. insulin lispro (Humalog®), insulin aspart (NovoLog®) or insulin glulisine (Apidra®), enable compensation of this specific defect. Weight gain during treatment with insulin lispro is somewhat lower than during treatment with regular insulin. Moreover, the rapid-acting insulin analogs offer the "usual" advantages like the possibility to inject the insulin after the meal or to correct hyperglycemia more quickly. Recently it was shown that meal-related own insulin production can be enhanced with rapid insulin releasers (medication class of **glinides**, repaglinide and nateglinide). You can consider such treatment alternatives only with appropriate basal replacement. Talk with your physician about which rapid-acting insulin and/or medication that suits your needs. Another hint: if you have problems with your weight, you might calculate and inject your insulin not per carbohydrate choices but per 100 kcal, independently of the carbohydrate content of the meal. That will effectively prevent eating of fat instead of "normal" food rich on carbohydrates.

In this case any tendency to eat only carbohydrate-free meals is thwarted, right?

Correct. Moreover, you will become more aware of how many calories you are consuming. Physical exercise and weight loss remain the "causal" and most effective therapy approaches for type 2 diabetes. Exercise generally improves the insulin action and reduces insulin resistance.

Oh, I know that already. Despite that, I am constantly overwhelmed by my weight problems. Can't I just take some medication to solve my weight problem?

Exercise and weight reduction are the only "causal" therapies against the metabolic syndrome

Actually, there are some initial signs of effective drug therapy for weight loss. Please be aware that each "efficient" therapy has its own unfavorable side effects as well. Therefore, these pharmacological treatments can be used only to treat "serious" obesity, similarly to surgical treatment (e.g. reduction of stomach size, stomach /"gastric"/ banding and similar approaches).

What can be done for people who are "only" moderately overweight, with type 2 diabetes components?

The most important behavioral components for weight reduction are still (1) increased physical activity, (2) reduced fat (and/or carbohydrates) in your diet, (3) meals rich in fiber (vegetables) and more fluids, especially water. For an effective weight loss or weight control you need to consider adaptation of your treatment while "dieting".

(1) **Slow weight reduction** with conventional dietary restrictions requires calorie reduction to about 800-1000 kcal per day. Fat and alcohol need to be avoided if possible. It is necessary to reduce insulin dosages, particularly delayed-acting insulin by 30-50% when food intake is cut back.

(2) Special **Low Calorie Vegetable Days**. Such special days with exclusively vegetables & fruits on a few days each week with lots of fluid, especially water and salads, fruits, vegetables. Such special days are recommended once or twice a week. On such special days you eat very few carbs (less than 4-6 choices), mainly as fruits, foods rich on fibers, lettuce and vegetables, all divided into 4-5 portions.

On such special vegetable days, metformin and sulfonylureas need to be withdrawn and the total insulin dosage reduced by half. With FIT, the basal insulin dosage needs to be halved in the evening and rapid-acting insulin needs to be maintained to cover the lower carbohydrate calorie intake. People usually find it useful and easy to have about two such days per week to reduce weight.

(3) Modified **fasting, very-low-calorie-diet (VLCD)** or **low-calorie-diets (LCD)** are in fact modified forms of fasting where the exclusion of solid food leads to a reduced stomach size which soon leads to a lower hunger perception. People who find the long-term adjustment of their lifestyle to be difficult may be more willing to try such "radical" form of weight reduction. Such low-calorie "formula diets" are available in many drugstores. If you consider a liquid-diet, select please such a formula that is particularly rich on fibers (liquid fibers), with details on carbohydrate and calorie contents on its labeling. Otherwise you will not be able to choose the correct amount of insulin. The main task is to optimize your lifestyle and eating under medical supervision (such as our *"Slim'n'FIT"* course). Oral anti-diabetic agents need to be omitted with this type of diet. If insulin is used for basal, it needs to be reduced to one third of usual, or less. Increasing your physical activity, to as much as your physical health allows, is crucial. If you do not use your muscles during your weight loss program, your body will reduce this muscle mass first, and then replace it with fatty tissue causing future weight gain. And that is definitely not desired!

Does this mean that trying to lose weight without exercise is to be avoided?

Correct. Physical activity has priority. Talk about it with your doctor and discuss what forms of physical activity are most suitable and best for you. Have fun, and get FIT! The *only* natural form of therapy in type 2 diabetes is to exercise. It does not matter which kind of training. Endurance training = at least 30 minutes per day with some 70% of your maximum effort which usually corresponds to the heart frequency of about "180 – age = training frequency" (check the data with your physician to make sure you are on the right track). That should be combined with strength training. Through training you minimize the insulin resistance, gain in muscular mass, downsize your belly and improve your blood pressure and fats as well. By the way, this also prevents development of autonomic neuropathy, which is the main cause of erectile dysfunction. Thus, you extend and improve your (also sexual☺) life... So, please, do not say that exercising is painful.

18. Further Suggestions, Possible Mistakes, Misunderstandings and Last Words of Advice

So far we have discussed many things. However, at the moment, I am not quite sure, if I'll be able to do all the things I plan to do...

Let us try to divide the most important things from the least important. Admittedly, carrying out your own insulin treatment where you are solely responsible for your own health can be a burden or a benefit depending on your experience. Your "level of empowerment" relies very much on your own organizational capabilities to use FIT in the most practical way in everyday situations.

What does it really mean?

You don´t need to read fairy tales to know you can´t have everything in life. So, you need to concentrate on what is most important to you. Our FIT patients achieve an average HbA_{1c} value of somewhat higher than 7% (7,1% corresponding to a mean blood glucose of 157 mg/dl). In other words, this value is *better than* that one achieved in the world-known type 1 diabetes study, *DCCT*, in the intervention group. Most people using FIT have such very good metabolic control; but of course not *all* of them. When we compare the behavior of people with excellent and not so good metabolic control on FIT, we can draw some conclusions. We advise you to do the following:

1. Regular and frequent self-monitoring. Blood glucose measurements and any necessary corrections are essential and cannot be replaced by any other means at this time. Do not skip a day. You need to do your self-checks even when you're having a "bad" or "stressful" day. Essential are at least four measurements per day in correlation with necessary corrections, or even 6 or more for the "labile" ones.

2. Write down the results of self-monitoring and insulin dosages. Patients who write down their results regularly in their log-book have much better HbA_{1c} values. You are likely to forget your therapeutic decisions or choices and blood glucose results if you do not write them down. Please consider your log-book as an essential part of your minimum equipment! Refer to the log-book at least 4 times a day. Record your insulin dosage or your blood glucose values shortly after taking them. To protect yourself from forgetfulness (*"Did I already inject the basal dose or did I only intend to?"*) or insecurity the log book is absolutely necessary. Anyone can neglect to fill in the log-book for a couple of days, but start again now! Don't forget that only a systematic adaptation of your algorithms will guarantee overall control, simple blood glucose corrections are not enough! **It is unlikely that you are able to adjust your algorithms correctly without written records**. For the same reason, it is necessary that you know your total daily insulin dose and calculate the daily mean blood glucose (MBG). MBG needs to stay above 100 (with the exception of pregnant women) and if possible under 160 mg/dl. For hypo-risk patients MBG should be somewhat higher (160-200 mg/dl). FIT record sheet for data acquisition can be downloaded straight from out web page: www.diabetesFIT.org or via www.lulu.com.

3. Updates, checkups and contacts.
Visit your diabetes physician or your FIT diabetes training and treatment center at least every three months, 4 times annually. If this is impossible because of distance, then make an appointment at least once or twice a year. We have developed a special refresher course: "*FIT Update*" especially for FIT-trained people. This training is very important when you need motivation to boost things back in order. *FIT Update* is a 2-day refresher course, a weekend event regularly offered and recommended to you about every 2-3 years. We try to motivate and keep you up-to-date with relevant knowledge and diabetes treatment advances. Never believe that you already know "everything" for your treatment so that you can abstain from a specialized, direct supervision where your specific situation can be discussed.

Never skip your periodic supervision sessions with your FIT & diabetes expert

4. Join diabetes self-help group or *FIT Alumni* support group. This is a good way to remain informed about what new diabetes treatments are available and how they can help you meet the demands of care. Emotional support from others in similar situations is a good reason to join a support group.

Self-help group:
People with
Insulin-dependent
diabetes

5. Always carry your minimal FIT equipment with you. But minimize the complexity of self-measurements and insulin shots to be able to perform them frequently enough. Enable yourself to carry out a more "functional" treatment.

Remember that disinfecting the skin is not necessary. Syringes with welded needles may be used several times. Blood glucose self-monitoring can be also done with visual strips if necessary (see www.betachek.com). Calorie counting is useful only for those who are striving to lose weight. If you have practiced estimating the quantity of carbohydrates in food, weighing food is not necessary. Once more: Always carry your minimal FIT kit with you.

6. Exercise. Why? Exercise improves the insulin effect and lessens its requirement. For your information: Type 1 diabetes (autoimmune disease) does not shield you from metabolic syndrome. This combination is undesired. Quite vital to know that exercise (30 min/day using 70% of maximal power) lessen the chance of coronary heart disease, lowers blood pressure, prevents from deterioration of brain functions due to aging, improves autonomic cardiovascular neuropathy (main cause of impotence) and reduces stress. It also makes you fit, sexy, less prone to depressions, and enhances weight reduction.

My blood glucose fluctuates greatly. Doctors tell me I have "brittle" diabetes.

Inappropriate use of insulin causes 99% of "abnormal" blood glucose fluctuations. You can avoid most of these fluctuations if you do the following:

1. Do not eat (or at least, don't eat "carbs") when your blood glucose is increased. No, you will not starve. Lower your blood glucose first and *only then* eat "carbs" when your blood glucose is within the target range.
2. Routinely adjust insulin absorption speed by appropriate measures (muscle injections, rubbing, heat, etc.) or use rapid acting insulin analogs instead of regular insulin for meals and corrections.
3. Do not inject in areas of the skin with "lipohypertrophy", an overgrowth of fat caused by injections in the same spot repeatedly. I know someone who had injected insulin in her thighs for 45 years. She had "cushions" there. She could not imagine other that areas of the body would be suitable for injections. Her blood glucose fluctuations improved dramatically as soon as she started to inject into another area.
4. Use better two injections of long acting insulin for basal replacement, even if only one (e.g. Lantus®) would potentially suffice. Make sure that your total delayed acting insulin does not exceed the half of your total daily insulin need.

What can I do when blood sugar fluctuations continue despite all these precautions?

Do not overemphasize sporadic high values. If you reach near normal HbA_{1c} values without serious hypoglycemia and you can simplify your life and eat freely, then you have fulfilled the aim of good control with FIT. If you lead a relatively regular life and follow prescheduled meals, exercise routines and injections, fluctuations will diminish (but probably with less freedom of lifestyle).

I am not sure if I can always keep my blood glucose under control.

It's not necessary when insulin is used sensibly and functionally. Take liberties with your "freedom" by using common sense and self-monitoring. You need comprehensive training in diabetes to control your blood glucose. Reading this book is not enough. Attending *FIT updates* boosts your motivation. As you already know, driving a car cannot be taught from the book alone; you need to put your knowledge into practice.

Do you think there will be more physicians and diabetes centers performing Functional Insulin Treatment in the future?

Yes, FIT USA Foundation was established years ago to disseminate FIT throughout the US. You can help by encouraging other people with diabetes to consider FIT and speaking to your physicians about your new treatment. FIT originated 25 years ago based on the experience of those affected by diabetes. You too can make a contribution to diabetes research and improve FIT by writing to me about your experience, ideas and potentially better solutions. I look forward to hearing from you! Please visit our web site (www.diabetesFIT.org) for more information on FIT.
I thank you in advance for all comments and improvements.

Vienna, Spring 2010

Professor Kinga Howorka, MD, MBA, MPH
kinga.howorka@meduniwien.ac.at
www.diabetesFIT.org

Acknowledgments

I would like to thank everyone whose suggestions led to the creation and the quality of this book. A special note of appreciation goes to members of the Research Group Functional Rehabilitation and Group Education, Vienna, Austria. In particular, RN Elsa Perneczky, Michaela Gabriel MD, Prof. Dr. Herwig Thoma, Susanne Reischl, Jiri Pumprla MD, MPH, MBA, and Doz. Andreas Thomas, as well as those who contributed suggestions in the international working group for FIT.

I am delighted to express my deepest gratitude and thanks to all members of FIT USA Foundation, particularly to Ms. Sheryl Hill, and to Professor Jay Skyler, Dr. Don Etzwiler, Professor Fran Kaufmann, who were so supportive during our FIT educational system transfer workshops held in the USA, and to many volunteers and cooperators.

Many thanks to Jane Speight, University of Holloway, London, for linguistic corrections of this book.

Zeljko Milak was extremely helpful preparing the “last” version of this manuscript – being probably the most annoying task of all. Thank you, Zeljko for your support.

Further, my special appreciation for many suggestions to early German editions to Professor Michael Berger, Professor Joszef Fövenyi, Dr. Monika Grüßer, Dr. Viktor Jörgens, Professor Ingrid Mühlhauser, Ms. Gertrude Reiss and Professor Werner Waldhäusl.

In the international training seminars for FIT from 1987-2009, altogether more than approx. 500 physicians and diabetes educators from Germany, Austria, Switzerland, Belgium, Czech Republic, Finland, Hungary, Poland, Romania and USA have been trained in the FIT method. If you are interested in International FIT training centers and activities of the international Research Group for FIT, please contact us at the ‘Research Group on Functional Rehabilitation and Group Education’, Medical University of Vienna, www.diabetesFIT.org.

Some Final Questions

Date:..

All questions have one or more correct answers.

A. Decide whether the following sentences are true or false.

	true	*false*
1. When fasting with type 1 diabetes and FIT, blood glucose values usually should stay between 60 and 90 mg/dl. This is a sign that the basal insulin on FIT is a correct dosage.	☐	☐
2. The "dawn phenomenon" is defined as the morning blood glucose drop in people with diabetes.	☐	☐
3. Blood glucose between meals will not decrease or increase with the correct basal rate during FIT.	☐	☐
4. The short acting insulin for eating and/or for blood glucose corrections can only be injected subcutaneously.	☐	☐
5. Blood glucose can be measured only with a blood glucose meter, which is regularly checked for accuracy.	☐	☐
6. On FIT, with an insulin requirement of 40-60 units/day, under basal conditions, 1 unit of rapid-acting insulin will induce a blood glucose drop of about 40 mg/dl.	☐	☐
7. If blood glucose values are under 160 mg/dl one hour after eating, this is the proof that the rapid-acting insulin dosage for the meal was accurate.	☐	☐
8. The adjustment of insulin absorption for prandial injections is seldom necessary with separate insulin injections of regular (not analog) insulin at every meal.	☐	☐
9. The optimal inject-eat interval before a breakfast rich in carbohydrates is 15 minutes for regular insulin when no acceleration of insulin absorption is induced.	☐	☐
10. In the case of increase of total insulin need (e.g. with infection), the algorithms for the basal and the prandial insulin should be increased to the same degree as the total daily insulin need has increased.	☐	☐
11. A basal rate established with long-acting insulin in the morning (Lantus®) and NPH or Levemir® insulin late in the evening is particularly recommended with high fasting values.	☐	☐
12. The lowest blood glucose values occur most often between 3 and 4 a.m.	☐	☐
13. The basal rate needs to be reduced by 10% in any case after an episode of serious hypoglycemia.	☐	☐
14. If delayed-acting and rapid-acting insulin are mixed in the same syringe, delayed acting insulin becomes rapid insulin.	☐	☐
15. A minimal equipment for FIT consists of blood glucose strips, rapid insulin, an insulin syringe, glucose tablets, record sheet and a pen.	☐	☐
16. Basal insulin is correct when you can fast for 36 hours without a decrease in blood glucose and without necessity to eat something to keep up with your basal.	☐	☐

B. FIT, you would like to play 2 hours of tennis in the afternoon. Which of the suggested solutions are correct?

	true	*false*
1. Reduce delayed-acting insulin; rapid insulin remains unchanged.	☐	☐
2. Eat 2-3 carbohydrate choices without prandial insulin.	☐	☐
3. Reduce regular and delayed-acting insulin by 2 units.	☐	☐

C. On FIT, you want to go cross-country skiing for a week. Which type of insulin adjustment do you suggest?

1. Reduce rapid insulin; delayed-acting insulin remains unchanged.	☐	☐
2. Reduce delayed-acting insulin; rapid insulin remains unchanged.	☐	☐
3. Reduce prandial and basal insulin at approximately the same proportion	☐	☐

D. A man with type 1 diabetes has been using FIT for a long time. He injects 25 U Lantus® at the evening. In addition, he takes 40-50 units of rapid insulin distributed throughout the day for eating and correcting hyperglycemia. His weight is "normal" and he eats approximately 14-18 carbohydrate choices per day. He is not happy with his fasting values of 140-240 mg/dl. Hypoglycemia is not a problem.

What insulin dosage algorithm changes are correct in this case?

1. Inject more Lantus® in the evenings.	☐	☐
2. Inject more Lantus® in the mornings and evenings.	☐	☐
3. Take the half of the basal dose as Lantus® in the morning and the other half as NPH or Levemir® late at the evening.	☐	☐

E. A person with diabetes who requires 45-50 units of insulin per day as total insulin consumption over 24 hours probably needs which of following sets of algorithms for insulin dosage?

1. Total basal insulin per 24 hours: 10 U. Prandial insulin for each carbohydrate choice: 1 U. Correction: 1 U of rapid insulin lowers blood sugar –30 mg/dl.	☐	☐
2. Total basal insulin per 24 hours: 20 U. Prandial insulin for each carbohydrate choice: 1.5 U. Correction: 1 U of rapid insulin lowers blood sugar –40 mg/dl.	☐	☐
3. Total basal insulin per 24 hours: 35 U. Prandial insulin for each carbohydrate choice: 3 U. Correction: 1 U of rapid insulin lowers blood sugar –15 mg/dl.	☐	☐

F. People at risk of severe hypoglycemia need above all to:

1. Drastically reduce the prandial and basal insulin dosage algorithms.	☐	☐
2. Choose a higher blood glucose target area and raise the algorithm: "1 unit of rapid insulin lowers my blood glucose by .....mg/dl."	☐	☐
3. Inject insulin no more than twice a day and eat carbohydrates every three hours.	☐	☐

Glossary

Acetone in the urine: see ketones
Acidosis: see ketoacidosis
Adrenalin: a stress hormone from the adrenal gland
Algorithms of insulin dosage: rules of insulin dosage. With FIT, they describe how much and which insulin is to be taken when fasting (basal) or eating (prandial) one carbohydrate unit, or for correcting blood glucose values beyond the current target.
Adjustment of insulin dosage, *primary*: with FIT, blood glucose correction with rapid acting insulin or with carbohydrates; an immediate reaction to a blood glucose value beyond the target area. ***Secondary* adjustment:** change of algorithms for the insulin dosage (basal, prandial or correctional), necessary when changing the average insulin requirement (e.g., when ill) or when glycemic control is insufficient.
Autonomic neuropathy: see neuropathy

Basal insulin: the insulin required with FIT to stabilize blood glucose with short-term fasting (between meals).
Beta cell: an insulin-producing cell in the "Langerhans" islands in the hormone-producing pancreas.
Blood glucose correction: deliberate lowering of blood glucose to the target, see adjustment of insulin dosage, primary.
'Brittle' diabetes: diabetes with "extraordinary" fluctuations in blood glucose. This unstable form occurs usually only in insulin-dependent diabetes mellitus.

Contraception: prevention of pregnancy.
Counter-regulatory hormones: stress hormones produced by the body (adrenalin, glucagon, cortisol, growth hormone), which increase blood glucose in opposition to insulin during hypoglycemia.
Creatinine: parameter of kidney function

Diabetic coma: ketoacidosis with loss of consciousness
Dawn phenomenon: increase of blood glucose in diabetes in the early morning, probably due to increased glucose production in the liver after an overnight fast.
Diabetic self-help groups: spontaneously organized groups of patients with diabetes mellitus to improve their knowledge and training in self-treatment and to share similar experiences.
Dialysis: artificial life-saving blood filtering, necessary with severe kidney damage

Erectile dysfunction (male impotence): disturbance of sexual function in men, lower ability to achieve or maintain an erection sufficient for satisfactory sexual intercourse. Most frequently caused by autonomic neuropathy in diabetes.
Estimated Average Glucose (eAG): Mean Blood Glucose derived from the HbA_{1c} value: $eAG = 28.7 \times HbA_{1c} - 46.7$

FIT, Functional Insulin Treatment: separate use of insulin either for eating or for fasting or for correction of high blood glucose.
Fluorescent angiography: a particular test of the eyes (retina), making the retinal

vessels visible, to diagnose diabetic retinopathy.

Glucagon: one of the counter-regulatory hormones produced in the alpha cells of the "Langerhans" islands (of the hormone-producing pancreas). Essential in the treatment of hypoglycemia with loss of consciousness; glucagon must be injected.

Glucose: simple sugar.

Glycemic control: actively maintaining blood glucose within a pre-set safe range. With FIT, the patient is actively involved in glycemic control based on primary and secondary adjustment of insulin dosage.

Glycogen: carbohydrate reserves (mostly in the liver)

Glycosylation: The process of linking and bondage of glucose to substances in the body (see Hemoglobin A_{1C}). Abnormal glycosylation of human tissues in insufficiently treated diabetes is probably the most important mechanism for the occurrence of late complications of diabetes.

Hemoglobin A_{1C}: glycated hemoglobin, the red pigment of red blood cells irreversibly bound to glucose molecules. Measurement of HbA_{1C} reveals the average mean blood glucose level over the past eight weeks before the blood test, see estimated Average Glucose.

Human insulin: insulin identical with that from humans. Usually of biosynthetic (DNA engineering) origin.

Hypoglycemia: abnormally low blood glucose levels (below 60mg/dl). In diabetic patients, this results from an insulin overdose.

Hyperglycemia: abnormally high blood glucose levels.

Immune intervention: medical treatment to change the defense performance of the body. Necessary, e.g., to suppress rejection reactions after transplants.

Insulin: the only blood glucose lowering hormone produced in the beta cells of the pancreas. Insulin enables glucose to enter the cells.

Insulin analogues: changed insulin with particular qualities, e.g., particularly fast–or slow-acting insulin.

Insulin antibodies: antibodies produced by the body against insulin. The production of insulin antibodies against human insulin is less likely than with animal insulin.

Insulin kinetics: time-dependent course of insulin concentration in blood.

Insulin-dependent diabetes mellitus: type1 diabetes mellitus, an autoimmune disease in which eventually (almost) no insulin is produced in the body.

Interdisciplinary care of pregnant diabetic women: care in a special center by several specialists (diabetologists, obstetricians, neonatologists, eye doctors).

Ketoacidosis: the presence of excessive ketone bodies in the body, indicating with hyperglycemia a lack of insulin.

Ketones: end-product of breaking down fat in the body, indicating in insulin-dependent diabetes that the body is using fat instead of glucose for energy. Usually associated with lack of insulin. If blood glucose is below 80 mg/dl, ketone bodies may also be a sign that food intake is too low.

Kidney threshold for glucose: indicates blood glucose values, at which, when reached the kidney begins to remove glucose into the urine.

'Langerhans" islands: clusters of hormone producing cells in the pancreas, among other the glucagon-secreting alpha cells and the insulin-secreting beta cells.

Laser treatment of retina: current therapy using a beam of light to retard the advancement of severe retinopathy.

Late complications of diabetes mellitus: characteristic changes in the capillaries (microangiopathy) and the larger blood vessels (macroangiopathy, atherosclerosis) as well as the nerves (neuropathy), induced by insufficiently treated long-term diabetes. The eyes and kidneys are the most affected organs.

Lipodystrophy: loss of fatty tissue, caused by long-term insulin administration in the same areas of the skin.

Long-acting insulin: contrary to regular or rapid acting insulin, insulin with delayed effects and prolonged action. Lantus® and Levemir® are long acting insulin analogs.

Macroangiopathy: a narrowing of the large blood vessels, atherosclerosis.

Microaneurysms: typical changes for diabetic retinopathy, formation of little "pouches" of capillaries of the retina; classical diabetes late complications.

Microangiopathy: a disturbance in the smallest blood vessels, known as capillaries, related to diabetes. Caused by changes in the capillaries, affects mainly the retina (eyes) and kidneys. Microangiopathy occurs usually first after 5-10 years of diabetes.

Microproteinuria, microalbuminuria: Excessive protein excretion in the urine. Mostly typical of diabetic nephropathy. Normal protein excretion rate in the urine is below 15 ug/min.

Nephropathy, diabetic: a disease of the kidneys, caused above all by diabetic microangiopathy. Is accompanied by high blood pressure and excessive protein excretion in the urine and eventually by increase of urinary substances (creatinine) in the blood. In the early stages of diabetic nephropathy, lowering the blood pressure as well as normalizing blood glucose are important to postpone kidney failure.

Neuropathy, diabetic: a disease of the nerves, which mostly occurs after long diabetes duration with poor glycemic control. Neuropathy can affect the nerves in the limbs and extremities, deteriorate heart rate control, delay gastric emptying and causes erectile dysfunction.

Non-insulin-dependent diabetes: type 2 diabetes in which the body is resistant to insulin. Also known as adult-onset diabetes.

Normoglycemia: normal blood glucose values: 60-120 mg/dl when fasting, up to 160 mg/dl after eating. The treatment target of diabetic persons with a long life expectancy.

Pancreas: an organ in the body that produces hormones like insulin and glucagon and is on the other hand involved in digestion.

Prandial: connected to meals or eating. Pre-prandial is before the meal, postprandial means after the meal.

Prandial insulin: the insulin necessary for transporting glucose from the food. With FIT, the prandial insulin is replaced by Regular or rapid acting insulin analogs. The prandial insulin dosage depends, most of all, on the quantity of carbohydrates in the meal.

Rapid acting insulin analogs: act two times quicker than regular insulin.

Regular insulin: insulin without delayed-acting substances, with a short-term effect:

rapid-acting insulin.

Retinopathy, diabetic: retinal disease associated with long term diabetes of 5-10 years duration.

Sensor: instrument for continuous measurement of certain parameters; e.g., a blood glucose sensor that continually measures the current blood glucose.

Speed of reabsorption of carbohydrates: speed of carbohydrate intake from stomach-bowel tract into the blood after a meal.

Speed of reabsorption of insulin: course in insulin concentration in blood after an injection.

Ultralente: long-acting insulin; delaying substances: zinc insulin crystals. Duration: from 18 to more than 30 hours (dependent on the dosage). This insulin preparation is not available any more.

Vitrectomy: surgical removal of vitreous body after it bleeds into the eye (with advanced diabetic retinopathy) to preserve the vision.

Answers to Questions

a) about basic diabetes training (Chapter 4)

A

True	False
A1	A2
A3	A7
A4	A10
A5	A13
A6	
A8	
A9	
A11	
A12	

B

True	False
B1	B2
B3	B4

C

True	False
C2	C1
	C3

D

True	False
D2	D1
	D3

E

True	False
E1	
E2	
E3	

F

True	False
F4	F1
F5	F2
	F3, F6

G

True	False
G5	G1
	G2
	G3
	G4
	G6

H

True	False
H2	H1
	H3

I

True	False
I2	I1
	I3

J

True	False
J1	J2
J3	J4
J5	

K

True	False
K2	K1
	K3

L

True	False
L1	L2
L3	L4
L9	L5
L10	L6
L14	L7
L16	L8
L18	L11
L19	L12
L26	L13
L29	L15
L31	L17
L32	L20
L33	L21
L45	L22
	L23
	L24
	L25
	L27
	L28
	L30
	L34
	L35
	L36
	L37
	L38
	L39
	L40
	L41
	L42
	L43
	L44

Answer to questions

b) about FIT training (Chapter 5)

A

True	False
A2	A1
A3	A5
A4	A7
A6	A8
A14	A9
A17	A10
	A11
	A12
	A13
	A15
	A16

B

True	False
B3	B1
	B2

C

True	False
C3	C1
C4	C2
	C5

D

True	False
D1.a	D1.b
	D1.c
D2.b	D2.a
	D2.c

E

True	False
E1.c	E1.a
	E1.b
E2.c	E2.a
	E2.b

F

True	False
F1.b	F1.a
	F1.c
F2.b	F2.a
	F2.c

Answers to final questions

A

True	**False**
A3	A1
A6	A2
A10	A4
A11	A5
A12	A7
A15	A8
	A9
	A13
	A14
	A16

B

True	**False**
B2	B1
	B3

C

True	**False**
C3	C1
	C2

D

True	**False**
D3	D1
	D2

E

True	**False**
E2	E1
	E3

F

True	**False**
F2	F1 and F3

Figures

Insulin Game: 'Celebration Day' / 'The Yielding to Temptation'

Name:

Date:

Eat what you want.

Whilst you are doing this (and afterwards!) answer the following questions:

1. *Insulin dose:* Can I choose the correct insulin dosage for a given meal?

   ____

2. *Insulin kinetics:* Can I choose the most suitable mode of insulin administration and the appropriate interval between injecting and eating?

   ____

3. *Practical skills:* Have I got everything I need in my "mini-kit"? That is, am I equipped to correctly estimate my blood sugar (even perhaps without a glucometer) and to administer insulin when I am on my own?

   ____

Time	Blood glucose [mg/dl]	Insulin [units]	Carbohydrate choices / meal	Notes

Insulin Game: Determining the Kidney Threshold

Patient:	**Date**:	**Trainer**:

Technical requirements Fulfilled: **Yes** **No**

1. Glucose meter with adequate strips available? ☐ ☐
2. Ketones strips available? ☐ ☐
3. Urine glucose strips available? ☐ ☐
4. Sufficient fluid available (approx. 2litres, but no beer or milk)? ☐ ☐

In order to start the test, you need to be able to answer "yes" to all questions. If not, please carry out the test another time when all technical requirements can be fulfilled.

Phase 1: Fulfilling the prerequisites for kidney threshold determination. Achieving basal conditions **Yes** **No**

1. Last meal at least 4 hours ago? ☐ ☐
2. Last rapid acting insulin injection at least 4 hours ago? ☐ ☐
3. Blood sugar stable for at least 1 ½ hours? ☐ ☐
   (No upward or downward trend)
4. Urine sugar negative? ☐ ☐
5. Urine ketones negative? ☐ ☐

If you have answered "no" to one or more of these questions, the kidney threshold test will not give meaningful results. In this case you need to conduct the test on another day when you are able to achieve basal conditions. During Phase 1 you need to drink at least 1 liter (four 8 oz. glasses) of any calorie-free liquid (water, mineral water, tea, etc.)

Phase 2: Controlled raising of blood glucose (Questions 1 and 2)
Eat enough glucose to raise your blood glucose to about 250 mg/dl. Try to pass enough urine to measure urine sugar in 10-min intervals. (This is relatively easy if you have consumed at least 1 liter of fluid during Phase 1). Measure your blood sugar at the same time that you measure your urine sugar. Drink another ¼ to ½ liters (one to two 8 oz. glasses).

Phase 3: 'BG-Plateau'
Wait until your blood sugar has stabilized and does not drop spontaneously. Only then can you use your algorithms to calculate the amount of rapid-acting insulin you need to lower your blood sugar to about 100-110 mg/dl. Drink another ¼ to ½ liters (1–2 8 oz. glasses).

Phase 4: Controlled lowering of blood sugar with regular insulin
Inject the amount of rapid acting insulin you have calculated that will be necessary to lower your blood sugar to 100-110 mg/dl. In order to save time, use the methods you have learned to speed up the action of rapid-acting insulin. Answer Question 3. Continue to measure both urine and blood sugar in 15 minute intervals

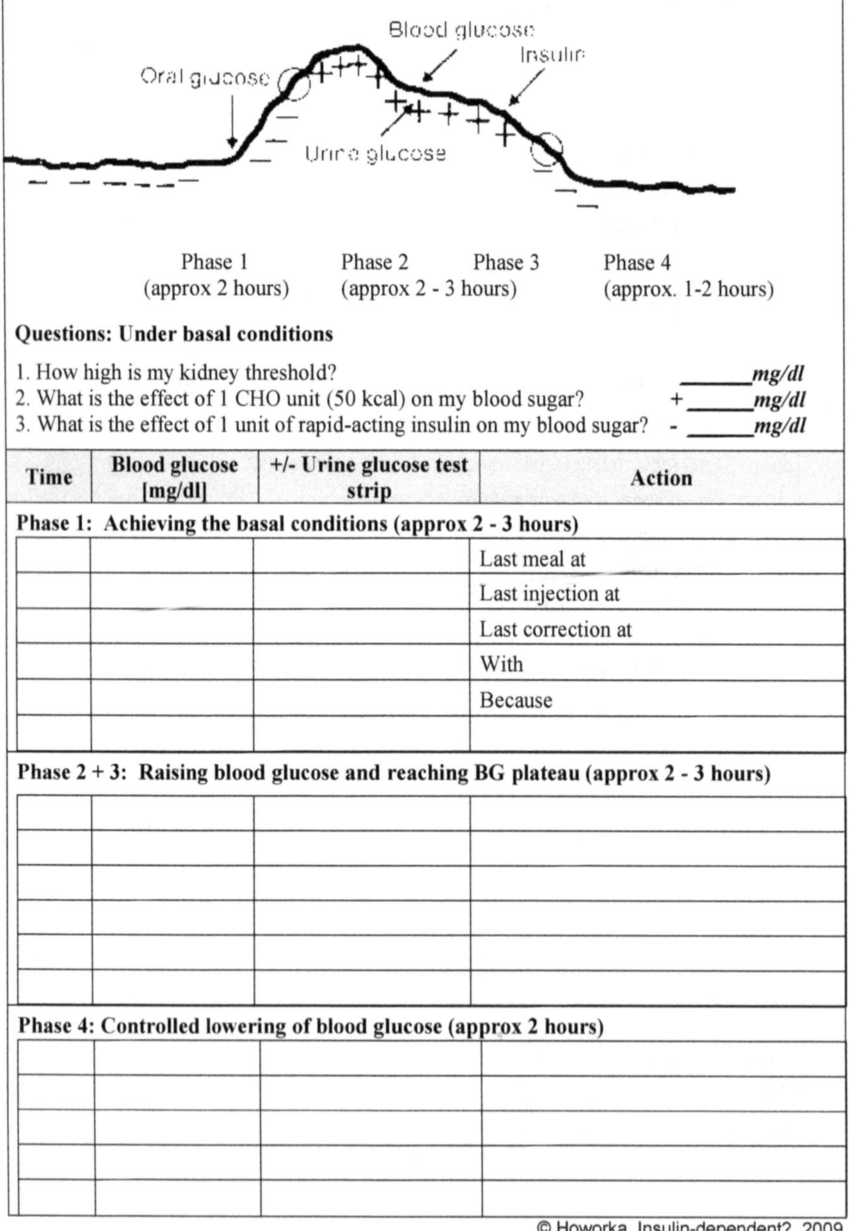

Phase 1 (approx 2 hours) | Phase 2 (approx 2 - 3 hours) | Phase 3 | Phase 4 (approx. 1-2 hours)

Questions: Under basal conditions

1. How high is my kidney threshold? ______*mg/dl*
2. What is the effect of 1 CHO unit (50 kcal) on my blood sugar? + ______*mg/dl*
3. What is the effect of 1 unit of rapid-acting insulin on my blood sugar? - ______*mg/dl*

Time	Blood glucose [mg/dl]	+/- Urine glucose test strip	Action
Phase 1: Achieving the basal conditions (approx 2 - 3 hours)			
			Last meal at
			Last injection at
			Last correction at
			With
			Because
Phase 2 + 3: Raising blood glucose and reaching BG plateau (approx 2 - 3 hours)			
Phase 4: Controlled lowering of blood glucose (approx 2 hours)			

Glycemic index		
Effect of chosen carbohydrate on blood glucose levels (BG effect of pure glucose =100%)		
Very high	90-100%	potatoes, honey, rice
High	50-90%	bread, muesli, also whole-wheat products
Low	30-50%	milk, pasta, fruit, ice cream, legumes
Very low	less than 30%	vegetables, beans, lentils

Table of Carbohydrate Choices

Attention: in the US "carbohydrate choices" include 15g of carbohydrates; portions are correspondingly larger.

One CHO Unit = 12 g carbohydrates (50 kcal from carbohydrates). Foods are listed in amounts containing 1 CHO unit. Those products which contain protein and/or fat also contain correspondingly more calories.

Bread and Cereals		
Wholemeal bread	25 g	1 small slice
White bread	20 g	1/2 thick slice
Roll	20 g	1/2 large roll
Crispbread, rye crisp	15 g	as indicated on package
Pasta	15 g uncooked,	1 tablespoon
	50 g cooked,	2—3 tablespoons
Rice	15 g uncooked,	1 level tablespoon
	50 g cooked,	2 level tablespoons
Flour	15 g	1 level tablespoon
Vegetables		
Potatoes boiled or roast	6o g	1 egg-sized potato
Lentils	20 g uncooked, 55 g cooked	2 tablespoons
Beans, peas, dried	20 g uncooked, 55 g cooked	2 tablespoons
Sweet corn	60 g cooked,	5 tablespoons
Fruit		
Apple	100 g	1 small
Orange	130 g	1 medium» without skin
Peach	120 g	1 medium
Banana	60 g 1/2 medium,	without skin
Pineapple	100 g	1 slice
Fruit juice (orange, pineapple)	25 g	1 small glass (4 oz.)
Milk Products		
Milk, whole or skimmed	250 g	1 cup (approx. 8 oz.)
Yoghurt	250 g	1 cup (approx. 8 oz.)

Conversion of mg/dl in mmol/l				
mg/dl	**mmol/l**		**mg/dl**	**mmol/l**
20	1,1		220	12,2
40	2,2		240	13,3
60	3,3		260	14,4
80	4,4		280	15,5
100	5,5		300	16,7
120	6,7		320	17,8
140	7,8		340	18,9
160	8,9		260	20,0
180	10,0		380	21,1
200	11,1		400	22,2

Conversion factor:
mg/dl x 0.05551 = mmol/l
mmol x 18.02 = mg/dl

Conversion of HbA1c in eAG (estimated Average Glucose)		
This table shows the correlation and conversion of HbA1c levels into an eAG level expressed in milligrams per deciliter (mg/dl) and mmol/l		
You can calculate eAG based on your HbA_{1C}: **eAG =28.7 x HbA1c –46.7 (mg/dl)**		
HbA1c %	**eAG mg/dl**	**eAG mmol/l**
6	126	7.0
6.5	140	7.8
7	154	8.6
7.5	169	9.4
8	183	10.2
8.5	197	11
9	212	11.8
9.5	226	12.6
10	240	13.4

(Modified after Nathan et al, 2008, and ADA; www.diabetes.org, download Sep2008)

Mathematical equations for initial algorithms are derived from our nomogram for initial FIT-algorithms (K. Howorka, Functional Insulin Treatment, Springer Publishers)
ATTENTION: Please ask your physician after the estimation of your algorithms whether these algorithms can apply for your particular case!

Algorithms of FIT	Short tests: "Insulin games"	Assessment criteria for every day
Basal insulin **How much (and which) insulin do I need even if I don't eat anything?** Total Daily Dose (TDD) x 0.45 = ______________ (of which 10-20% is used as "morning mound" + 80-90% as delayed acting insulin)	**Basal insulin** One-day-fasting (should require maximum 2-3 carbohydrate choices to keep up with basal rate without prandial insulin when fasting for 36 hours)	**Basal insulin** • blood glucose stability with short-term fasting (between meals) • fasting values (mostly between 90-140 mg/dl) • daily proportion (balance) delayed-acting insulin to short-acting insulin
Prandial insulin **How much short-acting insulin do I need for 1 carbohydrate choice?** TDD x 0.04 = _________ (average insulin need per 1 carb choice)	**Prandial insulin** So-called „yielding to temptation" or "celebration day" (correct if blood glucose in the target earlier and later after the meal)	**Prandial insulin** • blood glucose, short-term (1-2 h) after eating: information mostly about absorption rate of insulin and carbohydrates. Prandial insulin dosage still can not be eventually determined • only blood glucose late (4-6 h) after the meal allows reliable judgment of the insulin quantity used for this meal
Blood glucose correction algorithms **By how much *mg/dl* does 1U of short-acting insulin *lower* my blood glucose?** 1700 : TDD = _________ *BG lowering by 1 U insulin* **By how much *mg/dl* does 1 carbohydrate choice *increase* my blood glucose?** 90 – (body weight in kg x0,5) = *BG increase by 1 carb*	**Blood glucose correction algorithms** Checking the correction algorithms and determining the kidney threshold for glucose, as a "side-effect"	**Blood glucose correction algorithms** Judging of correction insulin dose only after completion of absorption, see above; individual risk of hypoglycemia determines target of blood glucose

Short test and all-day criteria for assessment of algorithms for insulin dosage for FIT. Note: TDD = Total Daily Dose, with usual eating habits and acceptable glycemic control (MBG=150-250 mg/dl). Calculation of BG lowering by 1U insulin applies *only* if proportion of basal insulin is below 40-50% of TDD!
A transfer to pump treatment will often lower total daily dose by 10-20%.

Index

Page numbers in bold refer to pages where the theme is treated as the main topics.

A

B

C

D

E

F

G

H

I

J

K

L

M

N

O

P

R

S

T

www.ingramcontent.com/pod-product-compliance
Lightning Source LLC
Chambersburg PA
CBHW060533210326
41519CB00014B/3212

* 9 7 8 1 4 4 6 1 8 7 5 3 1 *